RADIO AMATEUR NOVICE CLASS THEORY COURSE

by

MARTIN SCHWARTZ
Formerly Instructor at American Radio Institute

Published by
AMECO PUBLISHING CORP.
220 EAST JERICHO TURNPIKE
MINEOLA, N.Y. 11501

AMATEUR RADIO NOVICE CLASS THEORY COURSE

PRINTED 1986
All Material Up To Date

ISBN No. 0-912146-18-4

Library of Congress Catalog No. 81-68001

Printed in the United States of America

EXPLANATION OF COURSE

This course has been written for the purpose of preparing you for the FCC Amateur Novice Class examination. The course will enable you to understand the requirements of the Novice Class written examination as specified in the Federal Communications Commission Novice Class study guide. The complete FCC study guide is given in Appendix 1 at the back of this book.

The course is divided into three sections. The first section consists of three lessons on basic DC and AC theory. Some of the AC theory is not required for the amateur examination, but is given to provide you with a good theoretical background in order to understand the lessons that follow. Section 2 discusses vacuum tubes, solid state diodes, transistors, and amplifiers. Section 3 covers transmitters, receivers, antennas, FCC rules and regulations, and amateur station operating practices.

There are a number of practice questions at the end of each lesson. These questions will enable you to check your knowledge of the material in the lesson. A final "FCC-type" Novice examination appears on page 114. You should take this test before taking the FCC test. Most of the questions after each lesson and on the sample "FCC-type" examination are of the multiple choice type because this type of question is used exclusively by the FCC. The correct answers to these questions can be found in Appendix 4. Throughout this course you will find numerous schematic diagrams. You will not be required to draw diagrams on the actual FCC examination. However, you should know the schematic symbols for the various parts and you should be able to recognize what the various circuit diagrams are.

From time to time, the FCC adds new information that the prospective Novice Class operator should know. When this new material is too late to be included in the text of this book, it appears as "**ADDITIONAL INFORMATION**" on page 116 and in the Novice Class Addendum that comes with the book. This new information should be thoroughly mastered prior to taking the examination.

The Novice Class license examination consists of a code test, as well as the theory test. The Ameco Publishing Corp. publishes code courses on tape that prepare you for the FCC code examinations. See back cover.

The Novice theory examination consists of 20 questions that are taken from a pool of 200 questions. Ameco Cat.# 27-01 contains this pool of 200 questions together with the answers. It is suggested that you obtain a copy of Cat.# 27-01 after you have gone through this course. See back cover for details. The Ameco Publishing Corp. also publishes the actual test. It is Cat.# EN-1 and can be purchased by the applicant and given to the Volunteer Examiner. See back cover for details.

GOOD LUCK!

TABLE OF CONTENTS

INTRODUCTION TO RADIO

Let us begin by defining communication as a means or system by which we exchange our thoughts, opinions, information and intelligence with others. We are all familiar with the various methods of communication in use today. These methods may be simple and direct or highly developed technically. For example, people engaged in conversation, either directly or by using a telephone, illustrate the most common and simple means of exchanging ideas. Or the system may be more complex, as in radio transmission and reception between two radio amateur operators.

Before the discovery and development of electricity and radio, people used simple and crude methods for transmitting intelligence. The early Indians used smoke signals and drum beats to convey messages from one tribe to another. Although these sound and sight systems of transmitting messages were adequate for early man, they proved to be more and more archaic as man moved upward on the ladder of civilization. As mankind progressed into modern times, the invention of the telegraph and telephone became milestones in the history of the progress of communication. The telegraph and telephone were then radically different from any previous communication system in that they used electrical devices for both the sender and the receiver, and a wire or cable as the medium for the transmission. It thus became possible to communicate between any two points on the face of the earth which could be bridged by a cable or wire.

The next significant stage in the progress of message transmission was the development of a system of communication called the WIRELESS. The wireless was superior to the telegraph and telephone since it used the air as a transmission medium rather than a wire or cable. Today, wireless transmission is known as RADIO COMMUNICATIONS.

In this course, you will study all of the technical aspects of a basic Radio Communications System so that you will be well equipped to operate your own radio transmitting station.

Let us, at this point, consider briefly a basic radio communications system as illustrated in block diagram form in Figure 1. The basic operation of this system is as follows: Someone speaks into the microphone which changes sound energy into electrical energy. This electrical energy is fed into the sender or TRANSMITTER. The transmitter generates electrical vibrations which, together with the energy output of the microphone, are fed to the transmitting antenna. The transmitting antenna radiates the electrical vibrations out into space in the form of electrical radio waves. These radio waves travel outward from the antenna in a manner similar to the outward motion of ripples from a central point of disturbance in a pool of water.

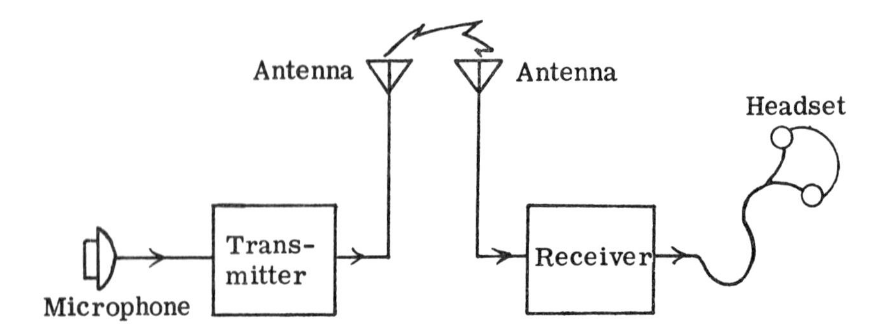

Figure 1. Block diagram of basic radio communications system.

At the receiving end of the radio communications system, the receiving antenna intercepts the radio waves and sends them into the receiver. The receiver converts the radio waves into electrical vibrations which energize the earphones. The earphones then convert the electrical energy back into the original sound that was spoken into the microphone attached to the transmitter. This brief description gives you a basic, non-technical picture of how a Radio Communications System operates.

SECTION I – LESSON 1
DIRECT CURRENT THEORY

1-1 MATTER AND ELECTRICITY

Matter is a general term used to describe all the material things about us. Matter includes all man-made structures, woods, metals, gases, etc.; in other words, everything tangible. All matter, regardless of its size, quality or quantity, can be broken down fundamentally into two different types of particles. These particles, which are too small to be seen under a powerful microscope, are called ELECTRONS and PROTONS. Electrically, we say that the ELECTRON is NEGATIVELY charged and the PROTON is POSITIVELY charged. Also, the proton is about 1800 times as heavy as the electron.

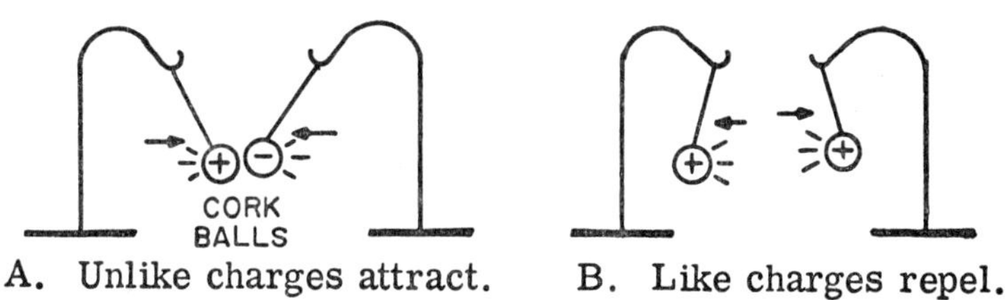

Figure 1-1. Attraction and repulsion.

1-2 THE LAW OF ELECTRIC CHARGES

Any object, such as a piece of glass, normally has a neutral or zero charge; that is, it contains as many electrons as protons. If this piece of glass can be made to have an excess of electrons, it is said to be negatively charged. Conversely, if the piece of glass can somehow be made to have a deficiency of electrons, the protons will predominate, and it is then said to be positively charged.

If a positively charged body is brought near a negatively charged body, the two objects will be drawn together. On the other hand, if two positively charged bodies or two negatively charged bodies are brought near each other, they will try to move away from each other. This reaction is the basis for our first law of electricity - The LAW OF ELECTRIC CHARGES. The law states "like charges repel, unlike charges attract". This law is illustrated by Figures 1-1A and 1-1B. In Figure 1-1A, a positively charged ball of cork is suspended by a piece of string near a negatively charged ball of cork. The two bodies swing toward each other since they attract each other. Figure 1-1B illustrates the two positively charged balls repelling each other.

1-3 DIFFERENCE OF POTENTIAL

If we were to connect a copper wire between the negative and the positive balls of cork, an electron flow would result. This is illustrated in Figure 1-2. The excess electrons from the negative ball would flow onto the positive ball where there is an electron deficiency and therefore, an attraction for the electrons. This flow continues until the deficiency and excess of electrons has disappeared and the balls become neutral or uncharged. This flow of electrons between the two differently charged bodies is caused by the difference in charge. A difference in charge between two objects will always result in the development of an electrical pressure between them. It is this electrical pressure that causes a current flow between these two bodies when they are connected by a piece of copper wire. This electrical pressure is defined as a DIFFERENCE OF POTENTIAL. The words "POTENTIAL" and "CHARGE" have similar meanings.

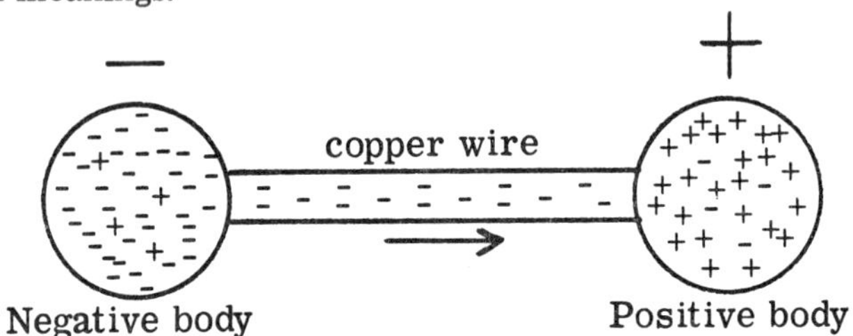

Figure 1-2. Flow of electrons.

1-4 CONDUCTORS AND INSULATORS

Materials through which current can easily flow are called CONDUCTORS. Most metals are good conductors. Conductors incorporate a large number of free electrons in their atomic structure. These free electrons are not held tightly and will move freely through the conductor when stimulated by external electrical pressure. Examples of good conductors, in the order of their conductivity, are silver, copper, aluminum and zinc.

Those materials through which current flows with difficulty are called INSULATORS. The electrons are tightly held in the atomic structure of an insulator and, therefore, cannot move about as freely as in conductors. Examples of insulators are wood, silk, glass and bakelite.

1-5 RESISTANCE

The ability of a material to oppose the flow of current is called RESISTANCE. All materials exhibit a certain amount of resistance to current flow. In order to compare the resistance of various materials, we require some standard unit of resistance measurement. The unit of resistance that was adapted for this purpose was the OHM, and the Greek letter Ω is its symbol. (For a list of common radio abbreviations, refer to Appendix 3). One ohm is defined as the amount of resistance inherent in 1,000 feet of # 10 copper wire. For example, 5,000 feet of # 10 copper wire would have a resistance of 5 ohms, 10,000 feet of # 10 copper wire would have 10 ohms, etc. Although the ohm is the basic unit, the MEGOHM, meaning 1,000,000 ohms, is frequently used. The instrument used to measure resistance is the OHMMETER.

1-6 RESISTORS

The resistor is a common radio part with a built-in specific amount of resistance. Resistors which are made of mixtures of carbon and clay are called carbon resistors. Carbon resistors are used in low power circuits. Wire wound resistors, which contain special resistance wire, are used in high power circuits. Figure 1-3 illustrates several types of fixed resistors together with the symbol which is used to represent them in circuit diagrams. When it becomes necessary to vary the amount of resistance in a circuit, we use ADJUSTABLE and VARIABLE resistors.

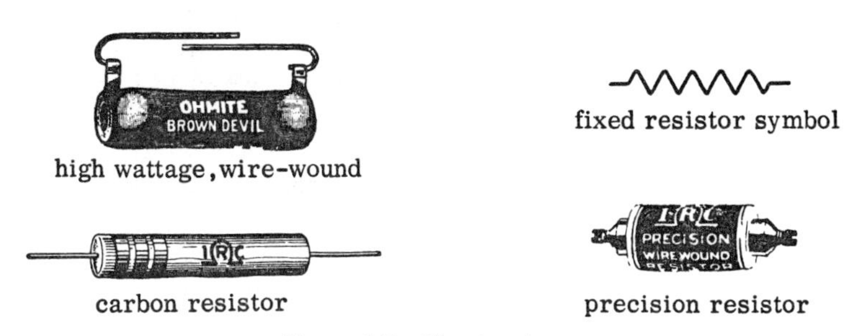

Figure 1-3. Fixed resistors.

The adjustable resistor is usually wire-wound, and has a sliding collar which may be moved along the resistance element to select any desired resistance value. It is then clamped in place. Figure 1-4A shows an adjustable resistor.

Variable resistors are used in a circuit when a resistance value must be changed frequently. Variable resistors are commonly called potentiometers or rheostats, depending upon their use. The resistance is varied by sliding a metal contact across the resistance in such a way so as to get different amounts of resistance. The volume control in a radio is a typical example of a variable resistor. Figure 1-4B shows a potentiometer used as a volume control for a radio receiver; Figure 1-4C shows a potentiometer wound of heavier wire for use in a power supply circuit.

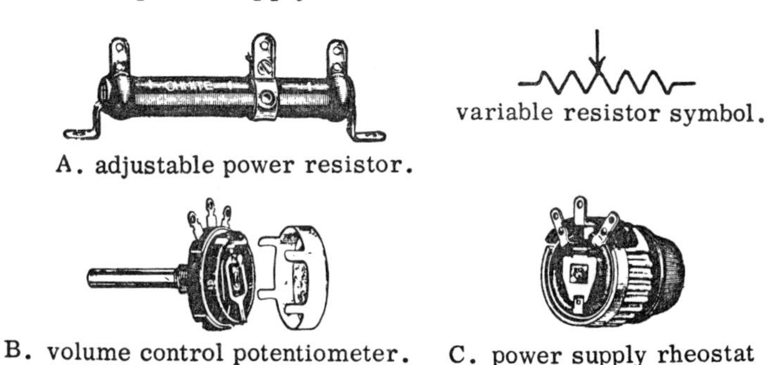

Figure 1-4. Variable resistors.

1-7 VOLTAGE AND CURRENT

Voltage is another term for the difference of potential or electrical pressure which we spoke about in a preceding paragraph. It is the force which pushes or forces electrons through a wire, just as water pressure forces water through a pipe. Some other terms used to denote voltage are EMF (electromotive force), IR DROP, VOLTAGE DROP, and FALL OF POTENTIAL.

The unit of voltage is the VOLT and the instrument used to measure voltage is the VOLTMETER. The KILOVOLT is equal to 1000 volts.

Current is the flow of electrons through a wire as a result of the application of a difference of potential. If a larger amount of electrons flow past a given point in a specified amount of time, we have a greater current flow. The unit of current is the AMPERE; it is equal to 6,300,000,000,000,000,000 electrons flowing past a point in one second. MILLIAMPERE and MICROAMPERE are terms used to denote one-thousandth and one-millionth of an ampere, respectively. Current is measured by an AMMETER. A meter whose scale is in milliamperes is called a MILLIAMMETER, and a meter that reads microamperes is called a MICROAMMETER.

The ammeter, voltmeter and ohmmeter, described above, can be combined into a single case, using one meter movement. This combination instrument is called a multimeter. A multimeter that can measure voltage, resistance, and current will suffice for most electronic servicing encountered outside of the laboratory.

1-8 THE DRY CELL

Several methods are used to produce current flow or electricity. A common method is the dry cell, which is used in the ordinary flashlight. The dry cell contains several chemicals combined to cause a chemical reaction which produces a voltage. The voltage produced by all dry cells, regardless of size, is 1½ volts. A battery is composed of a number of cells. Therefore, a battery may be 3 volts, 6 volts, 7½ volts, etc., depending upon the number of cells it contains. The fact that a cell is larger than another one indicates that the larger cell is capable of delivering current for a longer period of time than the smaller one. Figure 1-5 illustrates a typical 1½ volt cell and a 45 volt battery. The 45 volt battery contains 30 cells.

Every cell has a negative and a positive terminal. The electrons leave the cell at the negative terminal, flow through the circuit and return to the cell

Fig. 1-5A. 1½ v. flashlight cell.

Fig. 1-5B. 45 v. "B" battery.

at the positive terminal. This type of current flow is known as DIRECT CURRENT (DC). Direct current flows only in ONE direction.

1-9 ELECTRICAL CIRCUITS

Figure 1-6A is a diagram of a complete electrical circuit. The arrows indicate the direction of the current flow. As long as we can trace the current from the negative point of the cell, all around the circuit and back to the positive point, we have a complete circuit. The important thing to remember is that current will only flow through a complete circuit.

The necessary parts for a complete circuit are:

(1) A source of voltage - the dry cell in Figure 1-6A.

(2) Connecting leads - the copper wire conductors in Figure 1-6A.
(3) A load - the bulb in Figure 1-6A.

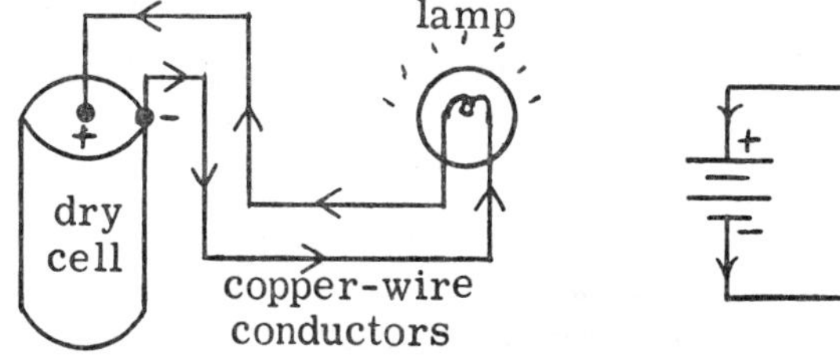

Fig. 1-6A. Complete electrical circuit.

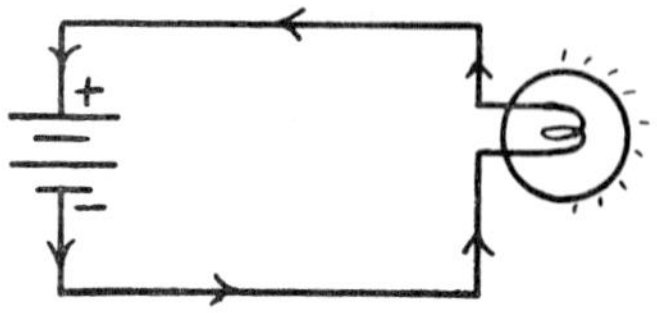

Fig. 1-6B. Schematic diagram.

If a break occurred in the connecting leads, or in the wire of the bulb, no current would flow and the bulb would go out. We would then have an OPEN CIRCUIT. Figure 1-7A illustrates the open circuit condition.

If we place a piece of wire directly across the two cell terminals, no current will flow through the bulb. This condition is illustrated in Figure 1-8A. The current by-passes the bulb and flows through the path of least resistance, which is the piece of wire. This condition is known as a SHORT CIRCUIT; it is to be avoided because it causes a severe current drain which rapidly wears the battery down.

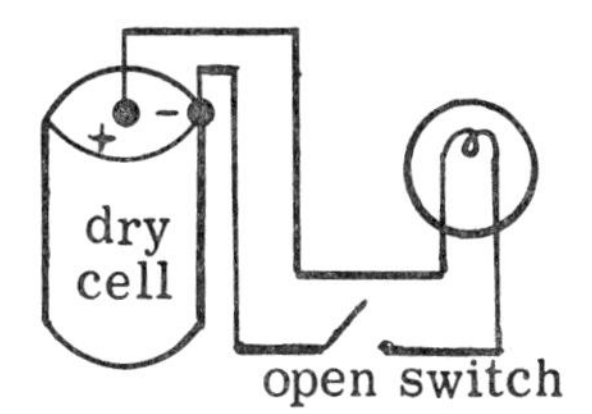

Fig. 1-7A. Open circuit.

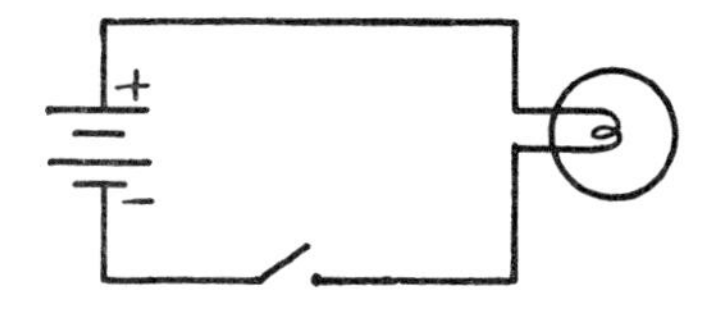

Fig. 1-7B. Schematic diagram.

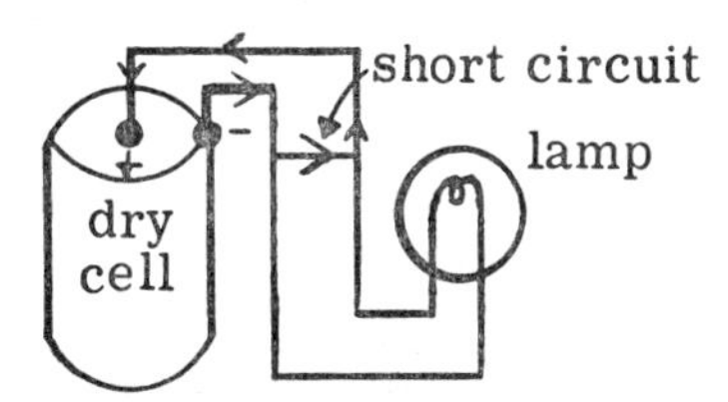

Fig. 1-8A. Short circuit.

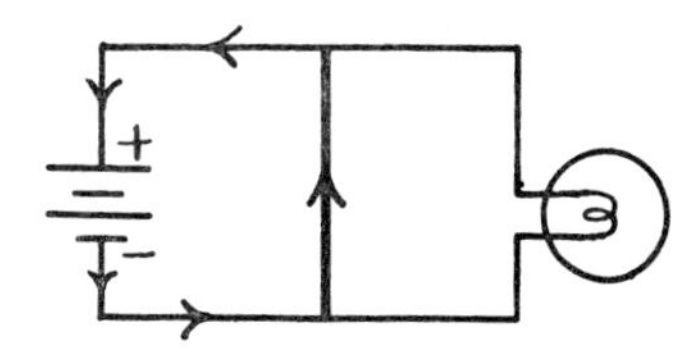

Fig. 1-8B. Schematic diagram.

1-10 SCHEMATICS

In drawing an electrical circuit on paper, we find it impractical to draw the actual battery or lamp as was done in Figures 1-6A, 1-7A and 1-8A. Instead, we use simple symbols to represent the various electrical parts. For instance:

A cell is shown as

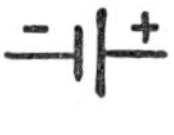

A battery is shown as

A resistor is shown as

You will find a complete table of radio symbols in Appendix 2. Figures 1-6A, 1-7A, and 1-8A can now be redrawn in the manner shown in Figures 1-6B, 1-7B, and 1-8B. Note that we indicate the negative battery terminal by a short line and the positive terminal by a long line.

1-11 OHM'S LAW

We have discussed the significance of voltage, current and resistance. Now we shall further study the important relationships that exist between these three factors. If we were to increase the battery voltage of Figure 1-6A, more electrons would flow through the circuit because of the greater electrical pressure exerted upon them. If we were to decrease the voltage, the flow of electrons would decrease. On the other hand, if the resistance of the circuit were made larger, the current would decrease because of greater opposition to current flow. If the resistance were made smaller, the current would increase because of smaller opposition. These relationships are formulated into a law, known as OHM'S LAW, which states as follows: The current is directly proportional to the voltage and inversely proportional to the resistance. Ohm's Law, mathematically stated, says that the current, in amperes, is equal to the voltage, in volts, divided by the resistance, in ohms.

The three formulas of Ohm's Law are:

(1-2) $I = \frac{E}{R}$ (1-3) $E = IR$ (1-4) $R = \frac{E}{I}$

"I" stands for the current in amperes.
"E" is the voltage in volts.
"R" is the resistance in ohms.

It is obvious that it is quicker to use letters such as I, E, and R, than to actually write out the words. Also, note that IR means I multiplied by R. If two out of the three factors of Ohm's Law are known (either E, I or R), the unknown third factor can be found by using one of the above three equations. Several examples will clarify the use of Ohm's Law:

PROBLEM: (1) Given: Current is .75 amperes
Resistance is 200 ohms
Find: The voltage of the battery.

SOLUTION: Since we are interested in finding the voltage, we use formula 1-3 because it tells us what the voltage is equal to. We then substitute the known values and solve the problem as follows:

(1) $E = IR$
(2) $E = .75 \times 200$
(3) $E = 150V.$

```
   200
 x .75
  1000
 1400
 150.00
```

PROBLEM: (2) Given: Battery voltage is 75 volts.
Resistance of bulb is 250 ohms.
Find: Current in circuit.

SOLUTION: Use formula 1-2 to find the current.

(1) $I = \frac{E}{R}$ (2) $I = \frac{75}{250}$ (3) I = .3 amp.

$$\begin{array}{r} .3 \\ 250 \overline{)75.0} \\ 75\ 0 \end{array}$$

PROBLEM: (3) Given: Current in circuit is 2 amp.
Battery voltage is 45 volts.
Find: Resistance of circuit.

SOLUTION: Use formula 1-4 and substitute for E and I to find R.

(1) $R = \frac{E}{I}$ (2) $R = \frac{45}{2}$ (3) R = 22.5 ohms.

1-12 RESISTANCES IN SERIES

If two or more resistances are connected end to end as shown in Figure 1-9A, any current flowing through one will also flow through the others. The

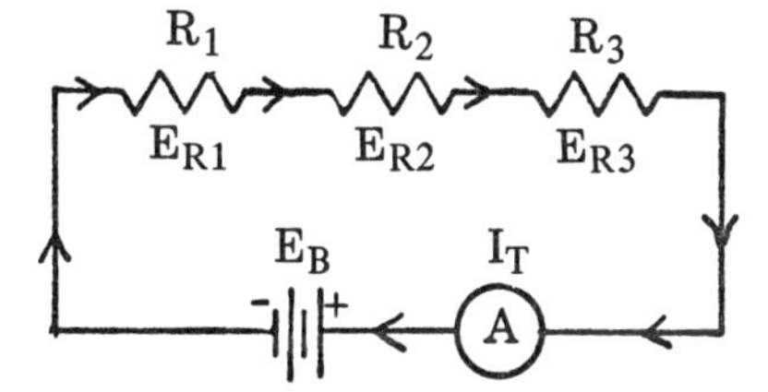

Figure 1-9A. Series circuit.

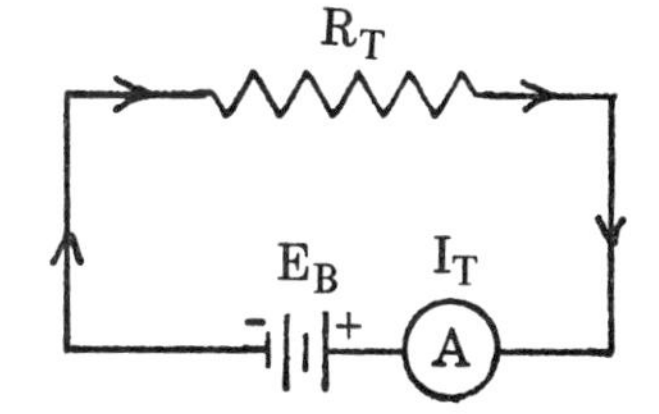

Figure 1-9B. Equivalent circuit.

arrows indicate the direction of current flow. The above circuit is called a SERIES CIRCUIT. Since the same current flows through each resistor, the CURRENT IS THE SAME AT EVERY POINT IN A SERIES CIRCUIT. Similarly, the total current is the same as the current in any part of the series circuit. To put it mathematically:

$$(1\text{-}5)\quad I_{Total} = I_{R1} = I_{R2} = I_{R3}$$

It is important to note that the current in Figure 1-9A will remain unchanged if the separate series resistors are replaced by a single resistor whose resistance value is equal to the sum of the three resistors. Figure 1-9b illustrates the equivalent circuit of Figure 1-9A.

THE TOTAL RESISTANCE IN A SERIES CIRCUIT IS EQUAL TO THE SUM OF THE INDIVIDUAL RESISTANCES.

$$(1\text{-}6)\quad R_T = R_1 + R_2 + R_3, \text{ etc.} \qquad \text{where } R_T \text{ is total resistance}$$

Whenever current flows through a resistance in a circuit, a part of the source voltage is used up in forcing the current to flow through the particular resistance. The voltage that is used up in this manner is known as the VOLTAGE DROP or fall of potential across that particular resistor. The voltage drop is equal to the current through the part multiplied by the resistance of the part.

If we add up the voltage drops across all the parts of a series circuit, the sum would be equal to the source or battery voltage.

$$(1\text{-}7)\quad E_B = E_{R1} + E_{R2} + E_{R3}, \text{ etc.}$$

where E_B is the battery voltage
E_{R1} is the voltage across R1
E_{R2} is the voltage across R2, etc.

PROBLEM: Find the resistance of R2 in Figure 1-9C.

SOLUTION: (1) Since we know the total current and the battery voltage, we can use Ohm's Law to find the total resistance.

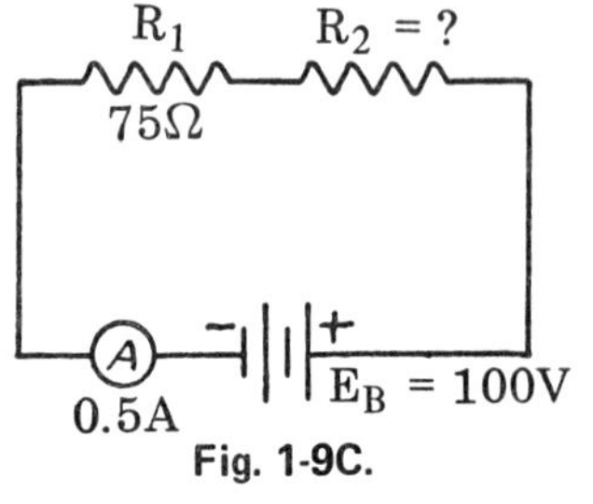

Fig. 1-9C.

$$R_T = \frac{E}{I} = \frac{100}{.5} = 200 \text{ ohms}$$

(2) Since the total resistance in this series circuit is 200Ω and $R_1 = 75\Omega$, then $R_2 = R_T - R_1$

(3) $R_2 = 200 - 75$ (4) $R_2 = 125$ ohms

PROBLEM: Find the voltage across R1 and R2 in Figure 1-9C.

SOLUTION: Since E = IR, the voltage across R1 is:

$$E_{R1} = .5 \times 75 = 37.5 \text{ V.}$$

The voltage across R2 is:

$$E_{R2} = .5 \times 125 = 62.5 \text{ V.}$$

Note that the total voltage divides itself across the resistors in proportion to the resistance of each resistor.

1-13 RESISTANCES IN PARALLEL

The circuit in Figure 1-10A is called a PARALLEL CIRCUIT. R_1 and R_2 are in parallel with each other. The current in the circuit now has two paths to flow through from the negative end of the battery to the positive end. If we remove resistor R_1 or R_2 from the circuit, the current has only one path to flow through from the negative to the positive end of the battery. Since it is easier for the current to flow through two paths instead of one, the total resistance of a parallel combination is less than the resistance of either resistor in the circuit. The more resistors we add in parallel, the less becomes the total resistance. This is because we increase the number of paths through which the current can flow. An analogy for this would be to consider the number of people that can pass through one door in a given time, compared to the number of people that can pass through several doors in the same time.

If the resistors in Figure 1-10A were equal, it would be twice as easy for the current to pass through the parallel combination than it would be for it to pass through either one of the resistors alone. The total parallel resistance would, therefore, be one-half of either one of the resistors. Figure 1-10B shows the equivalent circuit of Figure 1-10A.

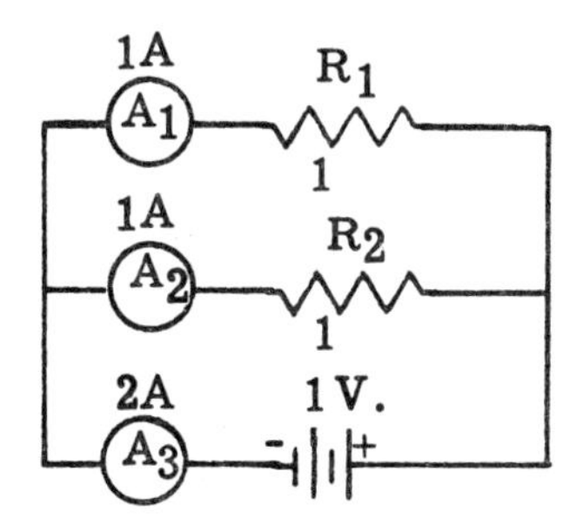

Fig. 1-10A. Parallel circuit.

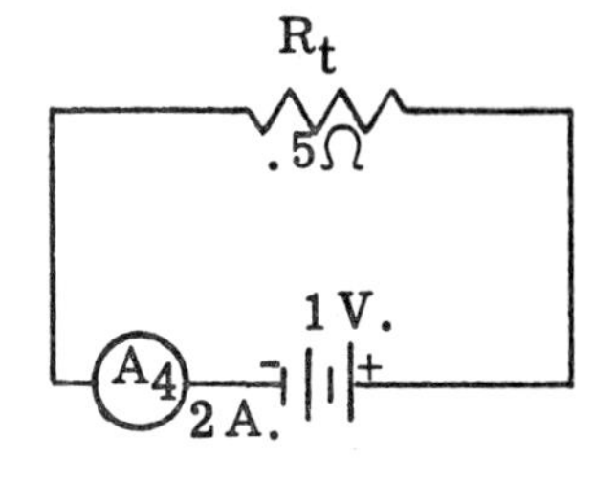

Fig. 1-10B. Equivalent Circuit.

In Figure 1-10B, R_T is the total resistance of R_1 and R_2 in parallel. The current flowing in the equivalent circuit must be equal to the total line current of Figure 1-10A.

The total resistance of ANY TWO resistors in parallel may be found by using the following formula:

$$(1\text{-}8)\quad R_T = \frac{R_1 \times R_2}{R_1 + R_2}$$

For example, if R_1 and R_2 of Figure 1-10A were 3 and 6 ohms respectively, the total resistance would be:

$$(1)\quad R_T = \frac{R_1 \times R_2}{R_1 + R_2} \qquad (2)\quad R_T = \frac{3 \times 6}{3 + 6} = \frac{18}{9} = 2 \text{ ohms.}$$

The total resistance of ANY NUMBER of resistors in parallel may be found by applying the following formula:

$$(1\text{-}9)\quad R_T = \frac{1}{\frac{1}{R_1} + \frac{1}{R_2} + \frac{1}{R_3} \text{ etc.}}$$

For example, if three resistors of 5, 10 and 20 ohms were connected in parallel, the total resistance would be:

$$(1)\quad R_T = \frac{1}{\frac{1}{R_1} + \frac{1}{R_2} + \frac{1}{R_3}} \qquad (2)\quad R_T = \frac{1}{\frac{1}{5} + \frac{1}{10} + \frac{1}{20}} \text{ (least common denominator is 20)}$$

$$(3)\quad \frac{1}{\frac{4 + 2 + 1}{20}} = \frac{1}{\frac{7}{20}} \qquad (4)\quad 1 \times \frac{20}{7} = 2\frac{6}{7} \text{ ohms.}$$

1-14 CHARACTERISTICS OF A PARALLEL CIRCUIT

1) The total resistance of several resistors hooked in parallel is less than the smallest resistor.

2) The amount of current flowing through each branch depends upon the resistance of the individual branch. The total current drawn from the battery is equal to the sum of the individual branch currents.

3) The voltage across all the branches of a parallel circuit is the same; in Figure 1-10A the voltage across R_1 is the same as the voltage across R_2.

A problem will illustrate the above principles. Refer to Figure 1-11.

GIVEN: Current through R_1 is 0.2A.
$R_1 = 50$ ohms. $R_2 = 200$ ohms.

FIND: 1) Current through R_2.
2) Total current.

SOLUTION: Since we know the resistance of R_1 and the current through R_1, we can find the voltage across R_1 by using Ohm's law.

(1) $E_{R1} = I_{R1} \times R_1$

(2) $E_{R1} = .2 \times 50$

(3) $E_{R1} = 10V$

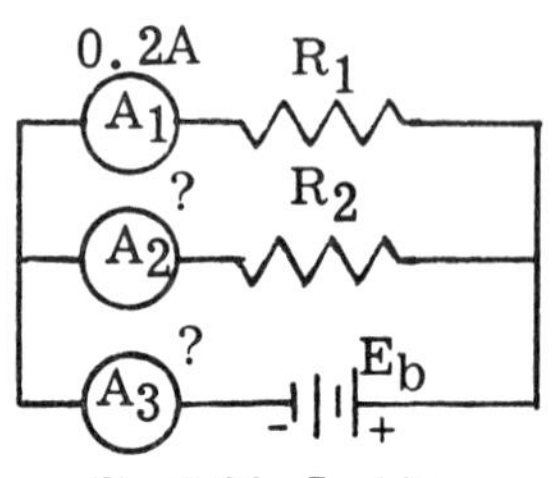

Fig. 1-11. Problem.

Since R_1 is in parallel with R_2, the voltage across R_2 is the same as that across R_1. Therefore, $E_{R2} = 10V$ also.

Knowing the resistance of R_2 (given) and the voltage across it, we can find the current through R_2:

$$I_{R2} = \frac{E_{R2}}{R_2} = \frac{10}{200} = .05 \text{ amp. current through } R_2$$

In a parallel circuit, the total current is equal to the sum of the individual branch currents; therefore:

(1) $I_T = I_{R1} + I_{R2}$ (2) $I_T = .2A + .05A = .25$ amp. total current

1-15 INTERNAL RESISTANCE OF A BATTERY

A battery has a certain amount of resistance, just as any other device has. We refer to the resistance of the battery as its "internal resistance". The current, flowing in a circuit, flows through the internal resistance of the battery in the same manner as it flows through the resistance of the load. The internal resistance of the battery is in series with the rest of the circuit. It is represented by "Ri" in Figure 1-12. The total resistance of Figure 1-12 is equal to Ri plus the load resistance.

If the battery is in good condition, Ri is small and is usually ignored.

Fig. 1-12

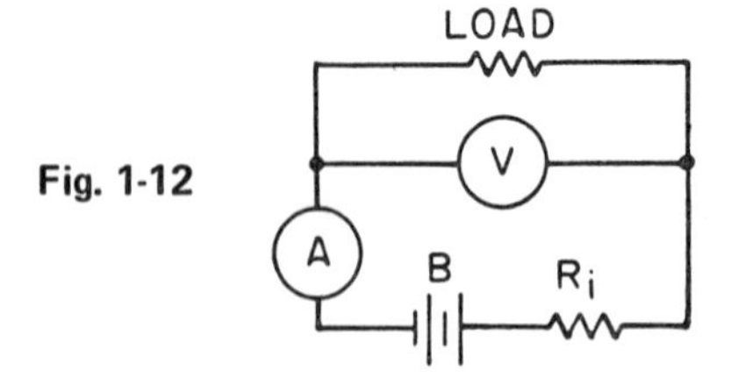

B - Battery.
Ri - Internal resistance of battery.
A - Ammeter.
V - Voltmeter.

1-16 POWER

Whenever current flows through a resistance, there is friction between the moving electrons and the molecules of the resistor. This friction causes heat to be generated, as does all friction. We could also say that electrical energy is changed to heat energy whenever current flows through a resistor. The unit of energy is the JOULE. The rate at which the heat energy is generated is the power that the resistor consumes. This power consumption in the form of heat represents a loss because we do not make use of the heat generated in radio circuits.

We should know how much heat power a resistor is consuming or dissipating. This is important because a resistor will burn up if it cannot stand the heat that is being generated by current flow. Resistors are, therefore, rated, not only in ohms, but in the amount of power that they can dissipate without overheating. The unit of electrical power is the WATT. A resistor rated at 5 watts is one which can safely dissipate up to 5 watts of power. If this resistor is forced to dissipate 10 watts, by increased current flow, it will burn up.

Exactly how much power is dissipated in a particular circuit, and upon what factors does the power dissipation depend? Since the power is the result of friction between the flowing electrons and the resistance in the circuit, the actual power dissipated depends upon the current and the resistance. The more current that flows, the more electrons there are to collide with the molecules of the resistance material. Also, the greater the resistance, the greater is the resulting friction. The actual power dissipated in a resistor can be found by the following formula:

(1-10) $P = I^2 \times R$ (I^2 means I x I)

where: P is the power in watts
I is the current in amperes
R is the resistance in ohms

PROBLEM: Find the power dissipated in a 2000 ohm resistor with 50 milliamperes flowing through it.

SOLUTION: First change milliamperes to amperes. This is done by moving the decimal three places to the left. Thus 50 milliamperes equals .05 ampere. Then substitute the values given in formula 1-10.

(1) $P = I^2 \times R$

(2) $P = .05 \times .05 \times 2000$

(3) $P = 5$ watts

$$\begin{array}{r} .05 \\ \times\ .05 \\ \hline .0025 \end{array} \qquad \begin{array}{r} .0025 \\ \times\ 2000 \\ \hline 5.0000 \end{array}$$

By using Ohm's law and algebraically substituting in formula 1-10, we can arrive at two more formulas for obtaining power dissipation.

(1-11) $P = E \times I$

(1-12) $P = \dfrac{E^2}{R}$

where: P is the power in watts,
E is the voltage in volts,
I is the current in amperes and
R is the resistance in ohms

Formula 1-11 states that the power is equal to the product of the voltage across the resistor and the current through the resistor.

The Wattmeter is the instrument that is used to measure power. The Watt-hour meter is the instrument that is used to measure energy.

1-17 FUSES

A fuse is a device that is used to protect the circuit from damage due to excessive current. It consists of a resistance unit with a fixed maximum current-carrying capacity. It is placed in series with the circuit or equipment that it is protecting. When the maximum current capacity of the fuse is exceeded, it burns up and opens the circuit. This protects the equipment in the unit from damage.

PRACTICE QUESTIONS — LESSON 1

(For answers, refer to Appendix 4)

1 The unit of power is the:
a. ampere b. coulomb c. watt d. joule

2 The instrument used to measure resistance is the
a. wattmeter b. ohmmeter c. ammeter d. voltmeter

3 The resistance of two equal resistors connected in parallel is:
a. the sum of the two resistors
b. one-half of one of the resistors
c. one-quarter of one of the resistors
d. the average value of the resistors

4 A kilovolt is:
a. 100 volts
b. one-thousandth of a volt
c. 1000 volts
d. one-millionth of a volt

5 The total current in a parellel circuit is:
a. the same in each branch
b. equal to the sum of the individual branch currents
c. equal to the current in each branch multiplied by two
d. none of the above

6 Of the following formulas, pick out the incorrect one:
a. $I = \frac{E}{R}$ b. $E = RI$ c. $R = \frac{I}{E}$ d. $P = I^2 R$

7 The total current in a series circuit is equal to:
a. the current in any part of the circuit
b. the sum of the currents in each part
c. the total resistance divided by the voltage
d. the sum of the IR drops

8 A short circuit:
a. is found in every good electrical circuit
b. causes a heavy current to be drained from the electrical source
c. prevents current from flowing
d. decreases the conductance of the circuit.

9 Find the power dissipated by a 2500 ohm resistor that is carrying 75 milliamperes.

10 Find the source voltage of the circuit shown:

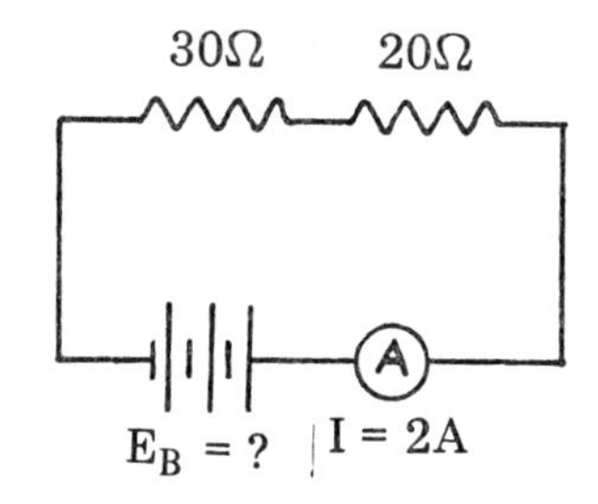

11 A 20 ohm resistor, a 15 ohm resistor and a 30 ohm resistor are all hooked in parallel. What is their total resistance?

12 The unit of energy is:
a. watt b. joule c. coulomb d. electron

13 The instrument used to measure power is the:
a. ohmmeter b. power meter c. wattmeter d. wavemeter

14 Another term for voltage is:
a. flow of electrons
b. electromotive force
c. EI drop
d. electrical energy

15 The unit of current is:
a. ampere b. coulomb c. joule d. ohm

16 The unit of electrical resistance is:
a. ampere b. volt c. ohm d. joule

17 The unit of electromotive force or potential difference is:
a. ampere b. volt c. watt d. coulomb

18 The instrument used to measure potential difference is:
a. wattmeter b. ammeter c. voltmeter d. ohmmeter

19 The total resistance of three 6 ohm resistors connected in series is:
a 18 ohms b. 9 ohms c. 2 ohms d. 3 ohms

20 The voltage drop in a 300 ohm resistor through which 5 ma. flow is:
a. 1500 volts b. 60 volts c. .017 volts d. 1.5 volts

SECTION I – LESSON 2
MAGNETISM

2-1 THE MAGNET

We are all familiar with the effects of magnetism. A horseshoe magnet will attract and pull to it iron filings. A powerful crane electromagnet will pick up heavy pieces of iron. A compass needle will point to the North Pole. A magnet, therefore, is any object which has the ability of attracting to itself, magnetic materials such as iron or steel. Figure 2-1 shows a horseshoe magnet attracting particles of iron filings.

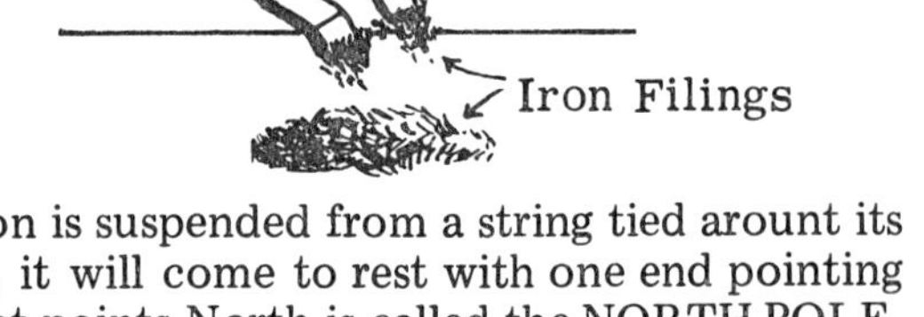

Figure 2-1.
Magnet's attraction power.

When a magnetized bar of iron is suspended from a string tied arount its center so that it is free to rotate, it will come to rest with one end pointing almost directly north. The end that points North is called the NORTH POLE, and the opposite end of the magnetized bar of iron is called the SOUTH POLE.

2-2 LAW OF MAGNETIC POLES

If the North Pole end of one magnet is brought near the North Pole end of another magnet, the magnets will repel each other. The same reaction of REPULSION will occur if two South Pole ends are brought close to each other. However, if a North Pole end and South Pole end are brought close to each other, the magnets will attract each other. The reason that the North Pole of a suspended magnet points to the earth's North geographical pole is that the earth itself is a magnet. The earth's South magnetic pole is located near the North geographical pole. A compass points to the North geographical pole because the compass needle is a magnet and the pointer is its North pole.

The results of experiments in magnetic attraction and repulsion were formulated into the law of poles which states that OPPOSITE POLES ATTRACT EACH OTHER, WHEREAS LIKE POLES REPEL EACH OTHER. Figure 2-2 illustrates this principle.

2-3 MAGNETIC LINES OF FORCE

We cannot see the forces of repulsion or attraction which exist between the pole pieces of two magnets. We must assume that the North Pole of one magnet sends out some kind of invisible force which has the ability to act through air and pull the South Pole of the other magnet to it. If we had unique vision, we would be able to see certain lines leaving the North Pole of one magnet and crossing over to the South Pole of the other magnet. These

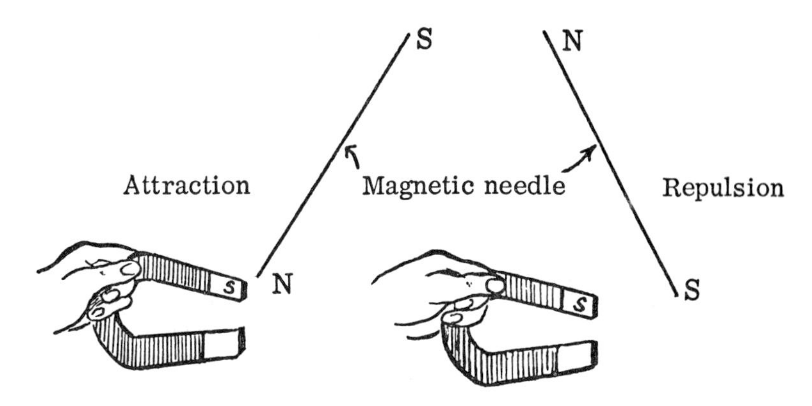

Figure 2-2. Attraction and repulsion.

lines are known as magnetic lines of force, and as a group are called a MAGNETIC FIELD or FLUX. Figure 2-3 illustrates the magnetic field as it exists around a bar magnet.

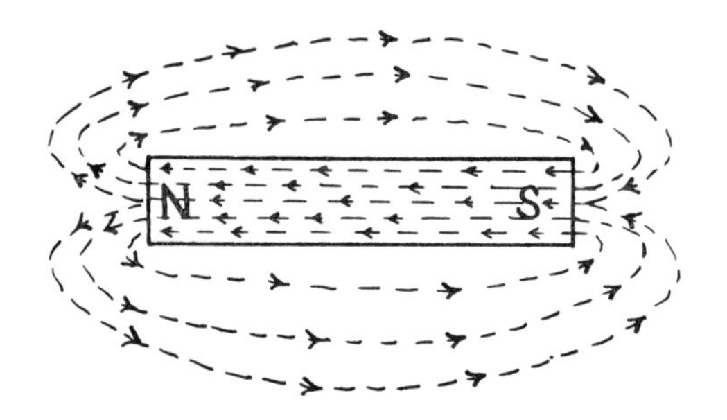

Fig. 2-3. Magnetic lines of force.

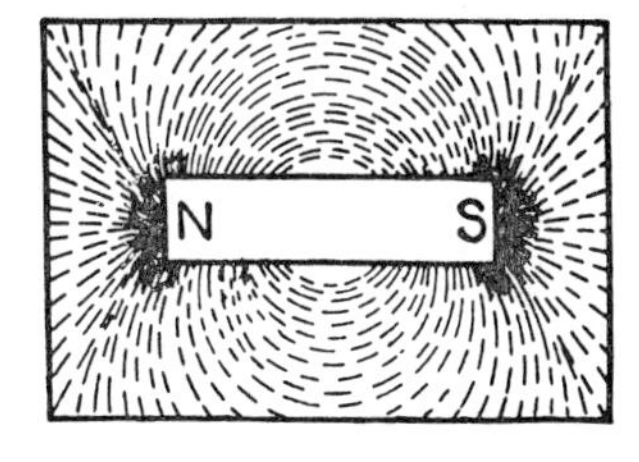

Fig. 2-4. Picture of iron filings.

Notice that the lines of force leave the magnet at the North Pole and return to the magnet at the South Pole. Note, also, that the magnetic field continues flowing inside the magnet from the South to the North Pole. The complete path of the magnetic flux is called the magnetic circuit.

One way to show magnetic lines of force is to sprinkle iron filings on a piece of paper under which we place a bar magnet. The result is shown in Figure 2-4. The iron filings arrange themselves so as to look like the lines of force that surround the magnet.

Figure 2-5 illustrates the magnetic field of attraction as it exists between the North and the South Pole of two separate magnets. Notice that the magnetic field appears to be actually pulling the two pole ends together.

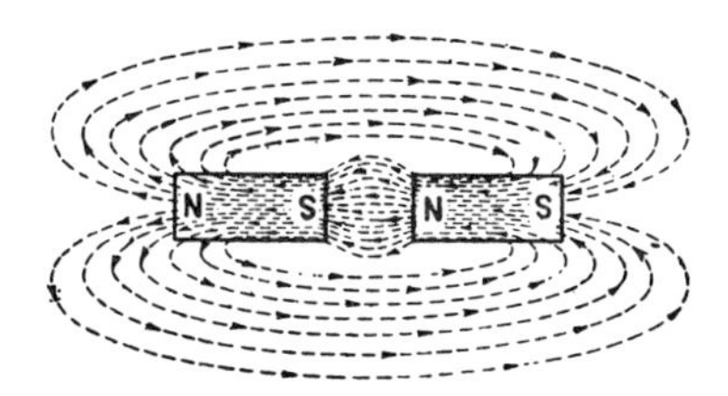

Fig. 2-5. Unlike poles attract.

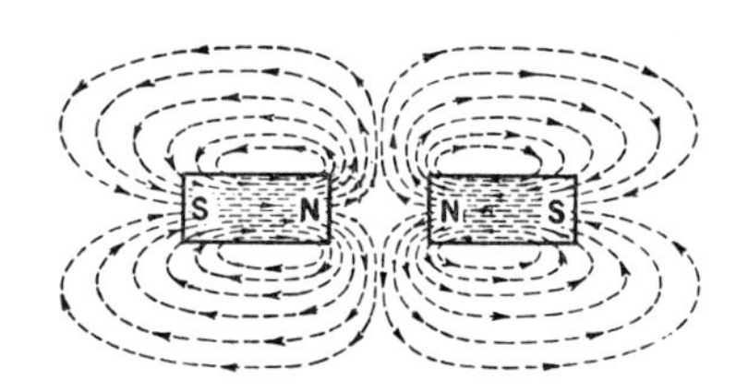

Fig. 2-6. Like magnetic poles repel.

Figure 2-6 illustrates the magnetic field of repulsion between two like poles. Notice that the magnetic fields are actually pushing each other away.

2-4 TEMPORARY AND PERMANENT MAGNETS

Soft iron can be magnetized easily by placing it in a magnetic field. However, as soon as the iron is removed from the magnetic field, it loses its magnetism. Such a magnet is called a TEMPORARY MAGNET. Steel or hard iron, on the other hand, which is difficult to magnetize, retains its magnetism after it has been removed from the magnetic field. A magnet of this type is called a PERMANENT MAGNET. Permanent magnets are usually made in the shape of a bar or a horseshoe. The horseshoe type has the stronger magnetic field because the magnetic poles are closer to each other. Horseshoe magnets are used in the construction of headphones and loudspeakers.

2-5 ELECTROMAGNETISM

The same type of magnetic field that we have been discussing exists around all wires carrying current. This can be proven by placing a compass next to a current-carrying conductor. It will be found that the compass needle will turn until it is at right angles to the conductor. Since a compass needle lines up in the direction of the magnetic field, the field must exist in a plane at right angles to the conductor. Figure 2-7 illustrates a current-carrying conductor with its associated magnetic field; the current flows from left to right and the magnetic field is in a counter-clockwise direction. In Figure

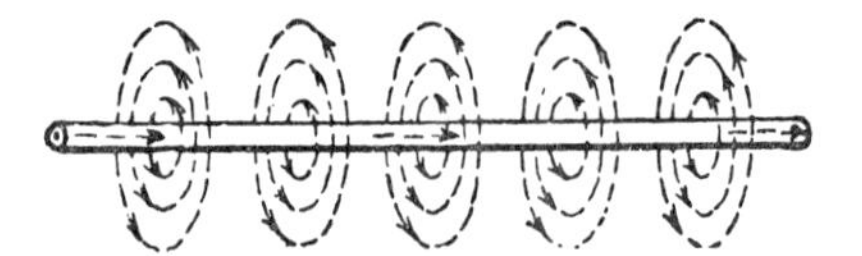

Fig. 2-7. Current left to right.

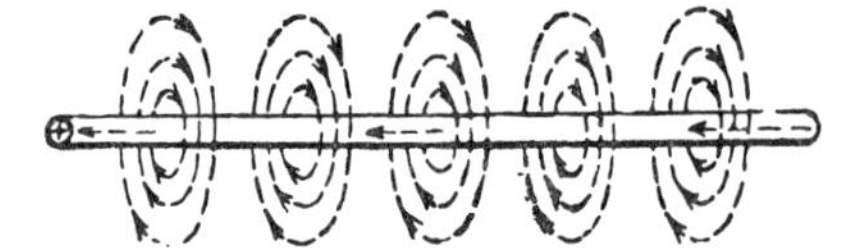

Fig. 2-8. Current right to left.

2-8, the current flows from right to left and the magnetic field is in a clockwise direction.

This magnetic field, of which only a number of cross-sections are shown, encircles the wire all along its length like a cylinder. Notice that the direction of the magnetic field, as indicated by the arrows, depends upon the direction of current flow in the wire.

2-6 THE COIL

If the same conductor is wound in the form of a coil, the total magnetic field about the coil will be greatly increased since the magnetic fields of each turn add up to make one resultant magnetic field. See Figure 2-9. The coil is called a SOLENOID or ELECTROMAGNET. The electromagnet has a North and South Pole, just like a permanent magnet. The rule for determining which end is the North Pole and which end is the South Pole states as follows: If we grasp the coil with the left hand so that the fingertips point in the direction of the current, the thumb will automatically point to the North Pole of the electromagnet. Thus, we see that the polarity of an electromagnet depends upon both the way in which the turns are wound and the direction of the current flow. If we reverse either the direction of the current flow or the direction of the windings, the North Pole will become the South Pole, and the South Pole will become the North Pole.

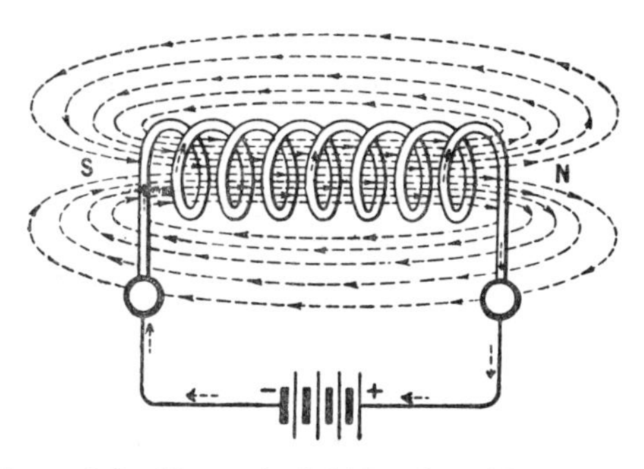

Figure 2-9. Magnetic field produced by current flowing through coil of wire or solenoid.

A compass placed within a coil carrying an electric current, will point to the North Pole of the coil. The reason for this is that the compass needle lines itself up in the direction of the magnetic lines of force. You will recall that inside a magnet, the direction of the field is from the South Pole to the North Pole. This is also true in an electromagnet, as illustrated in Figure 2-9.

There are various factors which influence the strength of an electromagnet. They are:

(1) The number of turns. An increase in the number of turns in a coil increases the magnetic strength of the coil.

(2) The amount of current. If we increase the amount of current in a coil, the magnetic strength increases.

(3) Permeability of the core. The core of the coil is the material within the coil. It may be air, glass, wood or metal. If we wind the coil on an iron core, we find that the strength of the electromagnet is increased by several hundred times over what it is with an air core. The iron is said to have more permeability than air; PERMEABILITY is the ability of a substance to conduct magnetic lines of force easily. If we have a core with a high permeability, we will have a large number of magnetic lines of force. This will result in a stronger magnetic field. Iron and permalloy are examples of materials having high permeability. Air is arbitrarily given a permeability of "one". The permeability of air is the basis for comparing the permeability of other materials. Iron and steel, for example, have a permeability of several hundred, depending upon the exact material.

Electromagnets are used in the manufacture of earphones, microphones, motors, etc.

2-7 MAGNETOMOTIVE FORCE

The magnetomotive force of a magnetic circuit is similar to the electromotive force of an electrical circuit. The magnetomotive force is the force which produces the magnetic lines of force or flux. The unit of magnetomotive force is the GILBERT. The number of gilberts in a circuit is equal to 1.26 x N x I, where N is the number of turns in the coil and I is the number of amperes. N x I, alone, is also known by the term AMPERE-TURNS. It is the number of turns multiplied by the number of amperes flowing in the circuit.

2-8 INDUCED VOLTAGE

If a coil of wire is made to cut a magnetic field, a voltage is induced in the coil of wire. The same reaction will occur if the magnetic field cuts the coil of wire. In other words, as long as there is relative motion between a conductor and a magnetic field, a voltage will be generated in the conductor. An induced voltage is sometimes called an induced EMF; EMF stands for electromotive force.

Figure 2-10A shows an iron bar magnet being thrust into a coil of wire. The dotted lines about the magnet represent magnetic lines of force. The relative movement between the coil and magnet will result in the turns of wire of the coil cutting the lines of force of the magnetic field. The net result of this action will be an induced voltage generated in the turns of the coil. This induced voltage will, in turn, cause a current to flow in the coil. A galvanometer (an instrument used to detect the presence of small currents) will deflect to the right indicating a current flow as a result of the induced EMF. Figure 2-10B shows the magnet being pulled out of the coil. The galvanometer needle will now deflect to the left indicating that the current is now in the opposite direction. Reversing the direction of the motion of the magnet in relation to the coil reverses the direction of the induced current as indicated by the position of the galvanometer needle and the polarity of the current flow.

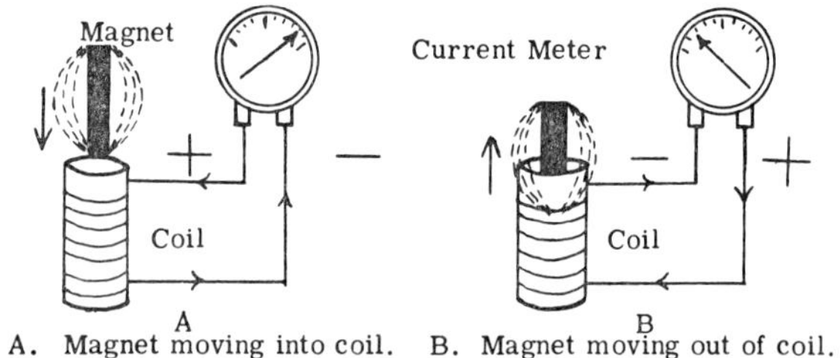

Figure 2-10. Inducing a voltage in a coil of wire.

This method of electromagnetic induction is used in the generators which supply us with our electricity. If we wish to increase the strength of the induced EMF, we can do the following:

(1) Use a stronger magnet.
(2) Use more turns on the coil.
(3) Move the magnet or the coil back and forth at a faster rate.
(4) Have the coil cut the lines of force at right angles if it is not already doing so. In other words, the more lines of force cut per second, the stronger is the resultant, induced EMF.

2-9 METERS

The most common meters such as the voltmeter, the ammeter and the ohmmeter, all make use of a basic meter movement known as the D'ARSONVAL type of meter movement. The D'Arsonval type of meter movement uses the principle of magnetic attraction and repulsion that has been described earlier in this chapter. A simplified illustration of the D'Arsonval movement is shown in Figure 2-11. A coil of fine wire is suspended by two spiral hair springs in a magnetic field created by a permanent horse-shoe magnet. A pointer is attached to the coil. If current flows through the coil, a magnetic field will be set up around the coil that will react with the field of the permanent magnet. This will cause the coil with its attached pointer to move. If we increase the current through the coil, the coil will move more. This is due to the increased magnetic reaction between the permanent magnet and the stronger field of the coil. When the current through the coil is removed, the coil will return to its original position by the springs that are attached to

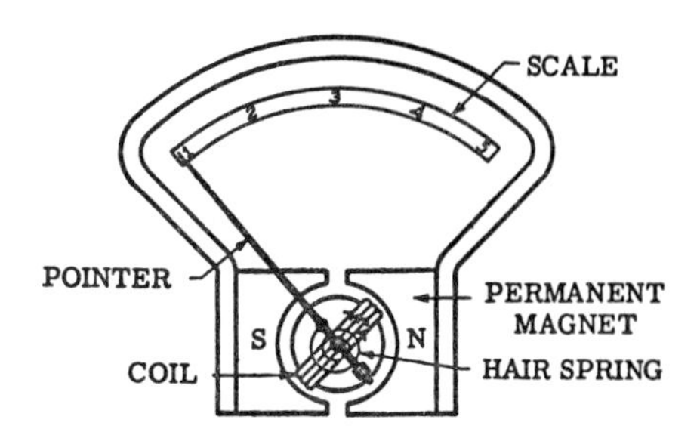

Fig. 2-11. The D'Arsonval meter movement.

the coil.

The pointer that is attached to the coil deflects across a scale, thereby indicating relative amounts of current that flow through the movement. If we add the proper scale and other parts to the basic D'Arsonval movement, we can convert it into a voltmeter, an ammeter, or an ohmmeter.

PRACTICE QUESTIONS — LESSON 2

1 The unit of magnetomotive force is:
  a. joule b. gilbert c. ohm d. rel

2 Permeability is:
  a. another name for magnetomotive force
  b. the ability of a coil to induce a voltage into another coil
  c. the ability of a material to conduct magnetic lines of force
  d. the ability of a magnet to retain the magnetism

3 If we placed a compass inside a coil carrying direct current, the North Pole of the compass would:
  a. point to the South Pole of the coil
  b. point to the North Pole of the coil
  c. point to the center of the coil
  d. shift back and forth until the current was shut off

4 Ampere-turns may be defined as:
  a. the square root of the number of turns multiplied by the current
  b. the number of turns multiplied by the square root of the current
  c. the number of turns multiplied by the current
  d. one-half the number of turns multiplied by the current

5 Inside of a bar magnet, the path of the lines of force is:
  a. from the North Pole to the South Pole
  b. from the South Pole to the North Pole
  c. either way, depending on the type of magnet
  d. there are no lines of force inside a magnet

6 The strength of an electromagnet will NOT be increased if we:
  a. increase the number of turns
  b. increase the permeability of the core
  c. change the iron core to an air core
  d. increase the current flow through the coil

7 What will happen as a result of a coil cutting a magnetic field?
  a. an increased magnetic field. c. an increase in the coil's inductance.
  b. an induced emf. d. an increase in the coil's capacitance.

SECTION I – LESSON 3
ALTERNATING CURRENT THEORY

3-1 INTRODUCTION

Up to this point, we have been studying DIRECT CURRENT (DC). Direct current flows in one direction only. We are now going to study a type of current which periodically reverses its direction of flow. This type is known as ALTERNATING CURRENT. A battery generates a direct current, and an alternating current generator generates alternating current. The abbreviation for alternating current is AC.

3-2 DEVELOPMENT OF THE ALTERNATING CURRENT WAVE

Figure 3-1 illustrates a loop of wire which can be rotated between the poles of a magnet. The magnetic field which exists in the space between the

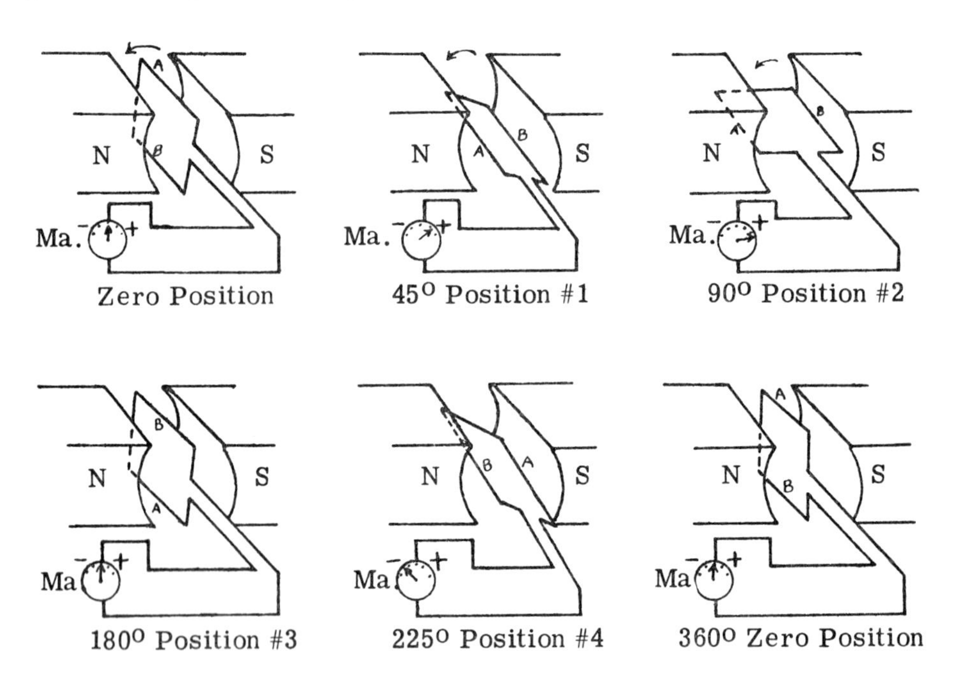

Figure 3-1. Generating the alternating current sine wave.

North and South Pole is not shown in the diagram. If the loop of wire is rotated through the magnetic field, an EMF (electromotive force) will be induced in the wires of the loop. This EMF will cause a current to flow in the circuit of the loop of wire. The milliammeter in series with the loop will indicate the current flow. From our previous study of magnetism, we know that an EMF will be induced in a conductor when it cuts through a magnetic field. One of the factors influencing the strength of the induced EMF is the relative cutting position of the loop as compared to the direction of the magnetic field. When the conductors of the loop cut perpencidular to the magnetic field, a maximum induced voltage will be generated. When the conductors of the loop are moving parallel to the magnetic field, no lines of force will be cut and therefore, no voltage will be generated. If the loop is rotated at a constant speed in a counter-clockwise direction, a current will flow whose strength and direction will vary with different positions of the loop. The strength and direction of the current for different loop positions is indicated in Figure 3-1. The resulting curve obtained is illustrated in Figure 3-2. At zero degrees, the loop begins its rotation with the ammeter indicating

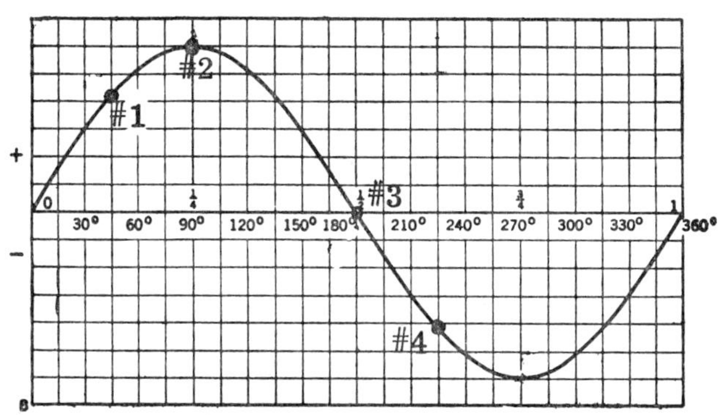

Figure 3-2. The sine wave.

zero current. (The conductors of the loop are moving parallel to the magnetic lines; therefore, no induced EMF will be generated). When the loop has reached position #1 (45 degrees), the current flow is indicated to be in a direction which we shall arbitrarily call positive; when the loop has reached position #2 (90 degrees), the current is at a maximum since the conductors are cutting into the magnetic field at right angles. The current flow is still in a positive direction. From position #2 to position #3, the current decreases in value and is still positive. At position #3 (180 degrees), the current is zero once again, as it was at the start. This is because the conductor is moving parallel with the magnetic field but is not actually cutting it. From position #3, through #4 and back to the starting position, the current goes through the same amplitude changes as it had gone through from starting position (zero degrees) to position #3 (180 degrees). However, from position #3 back to position zero, the direction of the current HAS REVERSED ITSELF and is now considered negative. The opposite to positive, or negative direction, is shown on the graph by drawing the curve below the horizontal line. The curve of Figure 3-2 representing the varying current through the loop, is a waveform known as an ALTERNATING CURRENT wave. The mathematical name for a fundamental alternating current wave is a SINE WAVE.

TO SUMMARIZE: Alternating current, as opposed to direct current, continuously varies in strength and periodically reverses its direction of flow.

3-3 CHARACTERISTICS OF THE SINE WAVE

An AC sine wave has the following important characteristics:

(1) The complete wave, as shown in Figure 3-2, is known as a CYCLE. The wave is generated in one complete revolution of the armature from 0 to

360 degrees.

(2) An alternation is one-half cycle, from 0° to 180°, or from 180° to 360°.

(3) THE FREQUENCY OF AN ALTERNATING CURRENT IS THE NUMBER OF COMPLETE CYCLES WHICH OCCUR IN ONE SECOND. If 60 such cycles were completed in one second, the frequency would be 60 cycles per second.

(4) The height of the wave at any point is known as its AMPLITUDE. The highest point of the wave is called the maximum or PEAK AMPLITUDE. In a sine wave, the peaks always occur at 90 degrees and 270 degrees; the zero points always occur at 0, 180 and 360 degrees.

3-4 FREQUENCY

The unit of frequency is cycles per second or simply cycles. The abbreviation for cycles per second is CPS or C. The meter used to measure frequency is called a FREQUENCY METER.

When a frequency is given as a large number of CPS, it can be converted into fewer units of kilocycles or megacycles, just as pennies can be converted into dollars. Example:

(a) 1000 cycles = 1 kilocycle. The abbreviation for kilocycle is "kc". (The prefix kilo always means 1000). Therefore, in order to convert cycles into kilocycles, we divide the number of cycles by 1000.

$$1{,}000{,}000 \text{ CPS} = \frac{1{,}000{,}000}{1000} = 1000 \text{ kc.}$$

(b) 1,000,000 CPS = 1 megacycle. (The abbreviation for megacycle is "Mc".) Therefore, in order to convert CPS into megacycles, we divide the number of CPS by 1,000,000.

$$1{,}000{,}000 \text{ CPS} = \frac{1{,}000{,}000}{1{,}000{,}000} = 1 \text{ Mc.}$$

In recent years, the term "Hertz" (abbreviated Hz), has been used in place of "cycles per second". Therefore, "kilohertz" (kHz) is used for kilocycles and "Megahertz" (MHz) is used for "megacycles".

The frequency of the AC power that is supplied to most homes in the United States is 60 Hertz. This is known as the POWER FREQUENCY. Radio waves transmitted by radio stations have a frequency much higher than the 60 Hz power frequency. Their frequency is generally above 20,000 Hz. An electrical frequency higher than 20,000 Hz is known as a RADIO FREQUENCY. The abbreviation for radio frequency is "RF". Figure 3-3 illustrates a low frequency of 60 Hz and an RF frequency of 1,000,000 Hz.

Sound waves which can be heard by the human ear are called audible sounds or audio sounds. The frequencies of the audio sounds lie in the range from 16 to 16,000 Hz. When a sound wave frequency is converted into an electrical frequency, it is known as an audio frequency (AF). For example, when our voice is amplified by a public address system, the sound waves from our throats strike the microphone and are converted into electrical frequencies, or audio frequencies.

3-5 INDUCTANCE

In Paragraph 2-6, we learned that a current-carrying coil of many turns

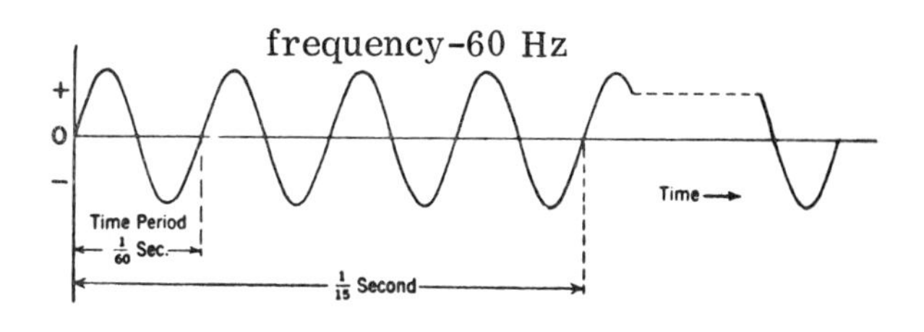

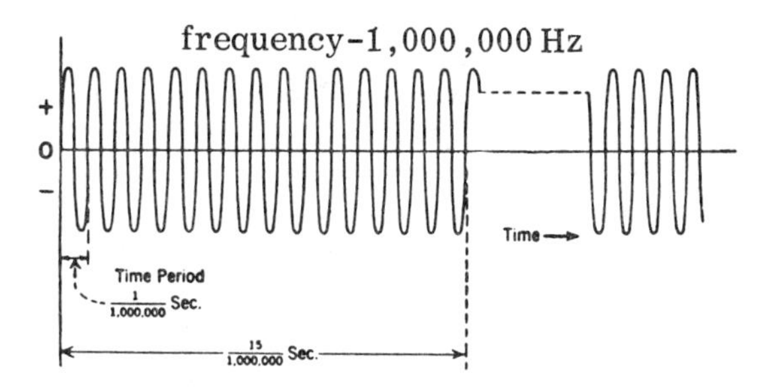

Figure 3-3. Low and high frequency wave.

behaves just like a magnet. The current will cause a magnetic field to surround the coil. If the current flowing through the coil is alternating, the magnetic field surrounding the coil will also be alternating. In Figure 3-4, we have a coil which has an alternating current flowing through it. This alternating current produces an alternating magnetic field around the coil which

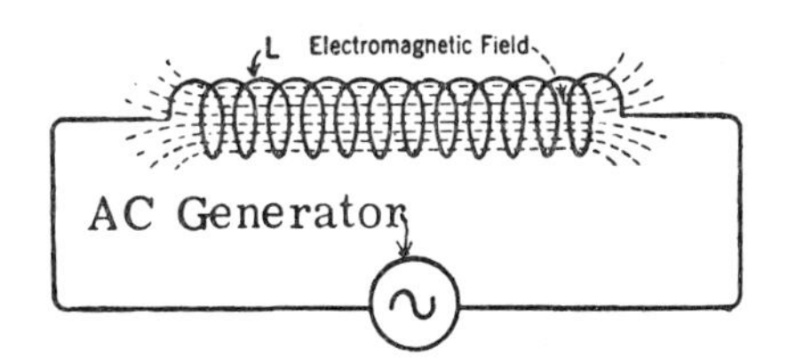

Figure 3-4. Coil with AC flowing through it.

expands and collapses in phase with the alternating current. When the current is zero, the magnetic field is zero; when the current reaches its peak, the magnetic field has reached its maximum value. Evidently, since the field starts from zero and builds up to a maximum, it is an expanding field. This expanding field must cut through the conductors of the coil itself. This cutting action induces an EMF in the coil which opposes the original current. The process wherein an induced EMF is generated in a coil which opposes the original current flow is called SELF-INDUCTION. The coil of wire is known as the INDUCTANCE. The unit of inductance is the HENRY, and the abbreviation of henry is h. The symbol for inductance is L. Smaller and more practical units of inductance are the millihenry (mh) and the microhenry (μh).

$$1 \text{ millihenry} = \frac{1}{1000} \text{ of a henry}$$

$$1 \text{ microhenry} = \frac{1}{1{,}000{,}000} \text{ of a henry}$$

The schematic symbol for inductance is:

or

Both types will be used throughout this course.

3-6 THE CAPACITOR

We have thus far studied two radio parts which exert a holding-back effect upon current: (1) resistors and (2) coils or inductors. We shall now investigate another holding-back device which has a tremendous application in radio; the CAPACITOR.

A capacitor is a device having, in its simplest form, two conducting plates separated from each other by an insulating material called a DIELECTRIC. The dielectric may be air, mica, oil, paraffined paper, etc. Figure 3-5 illustrates a two-plate capacitor connected across a battery. (Figure 3-7 shows the symbol for a capacitor.)

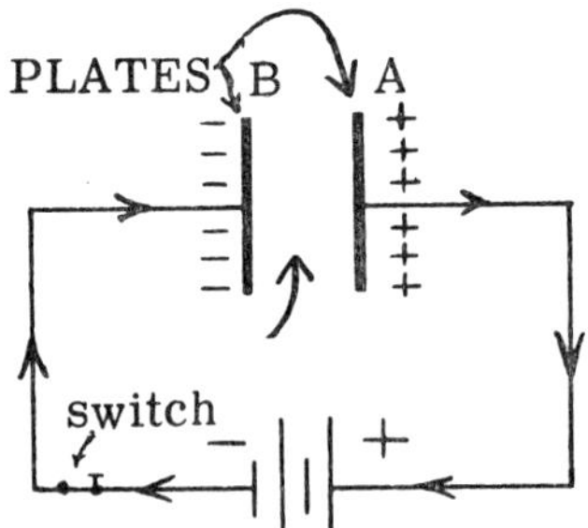

Fig. 3-5. Charging the capacitor.

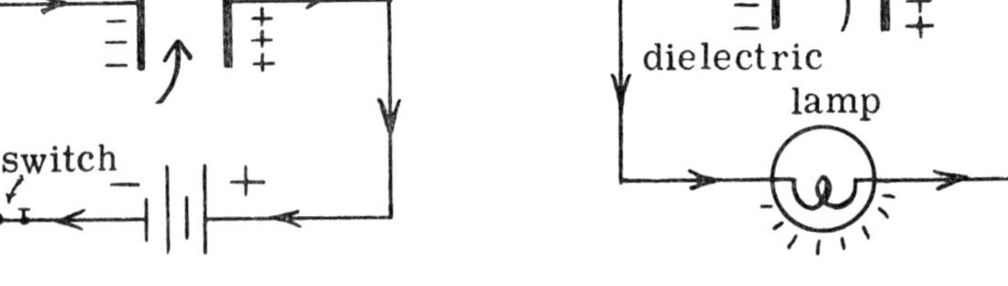

Fig. 3-6. Capacitor discharge.

When the switch is closed, a certain number of free electrons on plate A will be attracted to the positive side of the battery. Plate A is left with a positive charge; at the same time, plate B will have the same number of electrons pushed on to it by the negative side of the battery. This electron flow continues until a charge is built up on the capacitor plates, which develops a voltage equal to the battery voltage. The plates of the capacitor are now said to be electrically charged. The charge on the capacitor plates depends upon the size of the plates (the capacity), and the force of the battery (the EMF). Notice that the accumulated electrons on plate B cannot cross to the other plate because of the insulator dielectric in between.

When the capacitor has become fully charged, the voltage across the capacitor is equal to the battery voltage. If we disconnect the battery from the capacitor, the capacitor will continue to hold its charge. If a lamp is now connected across the charged capacitor (see Figure 3-6), the electrons on plate B will flow through the lamp and onto positive plate A where there is an attraction for them. During the brief duration of the electron flow, the lamp will light for an instant, indicating that a current has passed through it. The electrons will continue to flow until plate B no longer has a surplus of electrons. Plate B is then said to have a zero charge. Plate B is now neutral and, of course, plate A will have regained its electrons so that it is also neutral. The capacitor is now said to be DISCHARGED. A capacitor, then, is a device in which electricity may be stored for a period of time until it is ready for use.

A capacitor is a storage tank for electricity, just as a gallon jug, for example, is a storage place for water. If we force water into the jug under pressure, the amount of water that will go into the jug will be determined by the capacity of the jug and also, the pressure of force that the pump exerts on the water. Similarly, the amount of electricity that a capacitor will hold depends upon the same factors as apply to the water jug, namely, electrical pressure and capacity. The greater the capacity and the greater the pressure (voltage), the more electrons the capacitor will store up on its plates.

3-7 CAPACITANCE

The capacitance of a capacitor is determined by the size, shape, number and spacing of plates and the dielectric material. The symbol for capacitance is C. The unit of capacitance is the FARAD; the abbreviation for farad is FD. or F. Since the farad is an extremely large unit of capacitance, it is very rarely used. The smaller and more common units of capacitance are the MICROFARAD and the PICOFARAD (formerly called the micromicrofarad). The symbol for microfarad is μFD or μF and the symbol for picofarad is PFD or PF. Lower case letters such as μf and pf are also used.

$$1 \text{ microfarad} = \frac{1}{1,000,000} \text{ of a farad}$$

$$1 \text{ picofarad} = \frac{1}{1,000,000,000,000} \text{ of a farad}$$

The range of capacitance used in electronics may vary all the way from 2 pf up to 3000 μf.

3-8 THE VARIABLE CAPACITOR

Figure 3-8 shows the schematic symbol of a capacitor whose capacity can be varied. This capacitor is known as a variable capacitor and is used wherever the capacitance in a circuit must be continuously variable as, for example, tuning controls in radio receivers and transmitters.

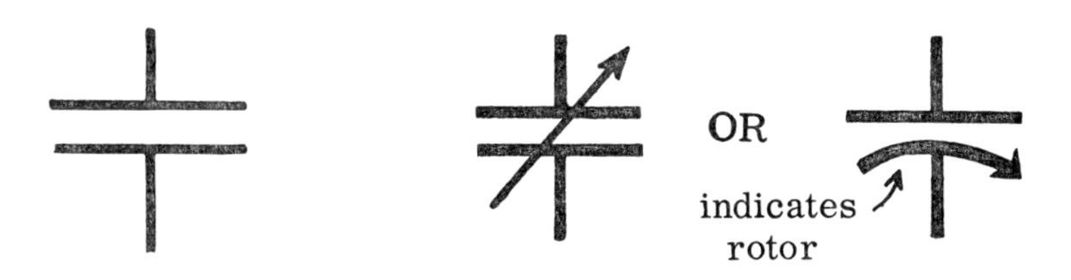

Figure 3-7.
Fixed capacitor symbol.

Figure 3-8.
Variable capacitor symbols.

Most variable capacitors are of the air dielectric type. A single variable capacitor consists of two sets of metal plates insulated from each other, and so arranged that one set of plates can be moved in relation to the other set. The stationary plates are the stator; the movable plates, the rotor, As the rotor is turned so that its plates mesh with the stator plates, the capcity increases. If several variable capacitors are connected on a common shaft so that all may be controlled at the same time, the result is known as a ganged capacitor. Figure 3-9 illustrates the rotor position of a variable capacitor for minimum, intermediate and maximum capacity.

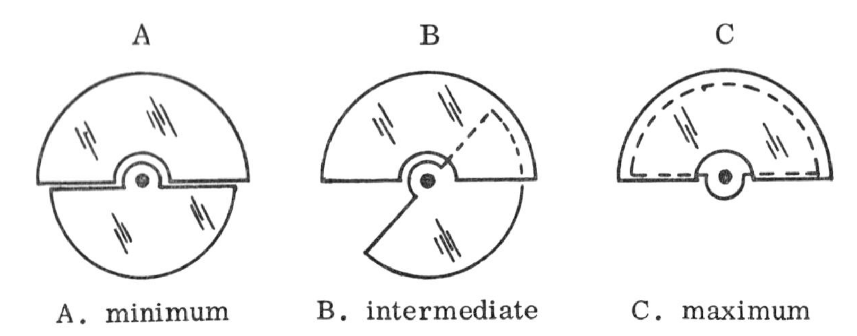

Figure 3-9. Variable capacitor settings.

3-9 RESONANCE

It has been stated that inductors and capacitors have a certain amount of opposition to alternating current. We call this opposition REACTANCE. The opposition that a capacitor offers to AC is called Capacitive Reactance (X_c) and an inductor's opposition is called Inductive Reactance (X_L).

The capacitive reactance that a capacitor offers to AC depends upon the value of the capacitor and the frequency of the current. Similarly, inductive reactance depends upon the value of the inductor and the frequency of the current. The unit of inductive and capacitive reactance is the OHM.

Fig. 3-10 shows an inductor and capacitor in series. If the frequency of the AC is such that the capacitive reactance is equal to the inductive reactance, we will have a condition known as RESONANCE. Since the two components are in series, we refer to the circuit as a SERIES RESONANT circuit. At resonance the two reactances cancel each other out and the current is at a peak.

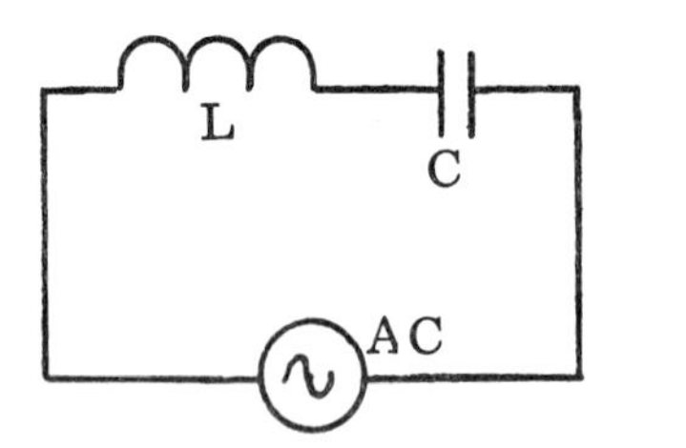

Fig. 3-10. Series Resonance

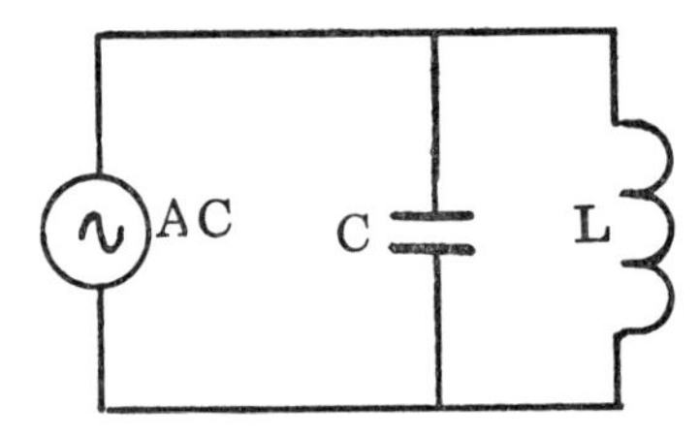

Fig. 3-11. Parallel Resonance.

For any given value of L and C, there is only one frequency that will cause resonance. This frequency is called the RESONANT FREQUENCY.

Figure 3-11 illustrates a PARALLEL RESONANT circuit. The circuits of Figures 3-10 and 3-11 are also referred to as TUNED circuits. These circuits are used in receivers and transmitters to select a particular frequency.

3-10 THE TRANSFORMER

The voltage supplied to most communities in the United States is the standard 117 volts AC. Many home radios require a voltage higher than 117 volts AC in order to operate satisfactorily. To fill this need, a device is incorporated in those radios to step up the line voltage of 117 volts to a higher voltage. The device which can increase or decrease the value of an AC voltage is known as a TRANSFORMER.

3-11 PRINCIPLE OF THE TRANSFORMER

You will recall from our early discussion of AC voltage that an EMF will be induced in a loop of wire which cuts into a magnetic field. As long as there is relative motion between the loop and the magnetic field, a voltage will be generated. If the loop is kept stationary and the magnetic field cuts across the loop of wire, the result obtained will be the same as if the loop were in motion instead of the magnetic field. In either case, a voltage will be induced in the conductors of the loop. The transformer operation is based upon a varying magnetic field inducing a voltage in a stationary coil of wire.

3-12 OPERATION OF THE TRANSFORMER

Every time current flows through a conductor, a magnetic field builds up around the conductor. The magnetic field is in phase with the current at all times. Therefore, if an alternating current flows through a coil of wire, an alternating magnetic field will exist about this coil. This alternating magnetic field expands outwardly away from the coil and collapses back into the coil periodically. If a second coil with a lamp across it is placed in the vicinity of coil # 1, as illustrated in Figure 3-12, the alternating magnetic field will cut across coil # 2 and induce an AC voltage in it; this will cause the lamp to

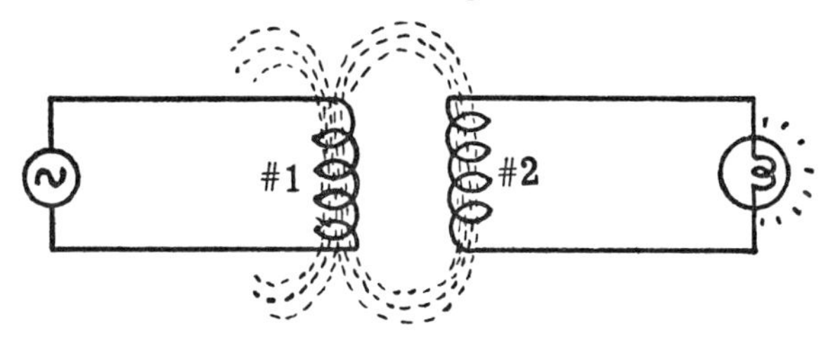

Figure 3-12. Magnetic coupling.

light. Notice that no electrical connection exists between the coils. Energy is transferred from coil # 1 to coil # 2 by means of the varying magnetic field. We say that the coils are MAGNETICALLY COUPLED. This method of

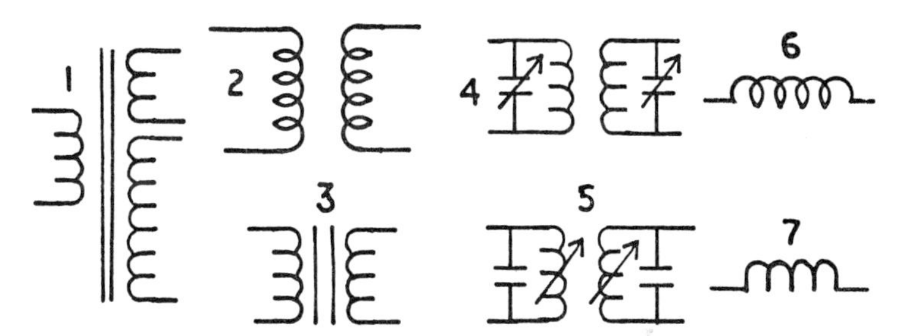

#1 - power transformer. #2 - RF transformer. #3 - audio transformer. #4 & #5 - IF transformers. Note that either symbol #6 or #7 may be used to represent coils of a transformer or choke.

Figure 3-13. Transformer symbols.

transferring energy from one coil to another is known as TRANSFORMER ACTION. The entire device consisting of two coils magnetically coupled is known as a TRANSFORMER. Coil #1, which is connected to the voltage source, is called the PRIMARY. Coil # 2, in which the induced voltage is developed, is called the SECONDARY.

The voltage relationship in a transformer depends directly upon the turns ratio of the transformer. If the secondary has twice as many turns as the primary, the secondary will have twice the voltage as the primary. If the secondary has half as many turns as the primary, it will have half the voltage of the primary. A transformer whose secondary has more voltage than the primary, is called a step-up transformer. If the secondary has less voltage than the primary, the transformer is called a step-down transformer.

Figure 3-13 shows the schematic symbols of typical transformers used in radio circuits.

3-13 THE POWER TRANSFORMER

The transformer in Figure 3-14B is known as an air-core transformer. Its use is confined to radio frequencies and it will be considered later on. A transformer which is used to transfer AC power at the power frequency of 60 Hz is known as a POWER TRANSFORMER.

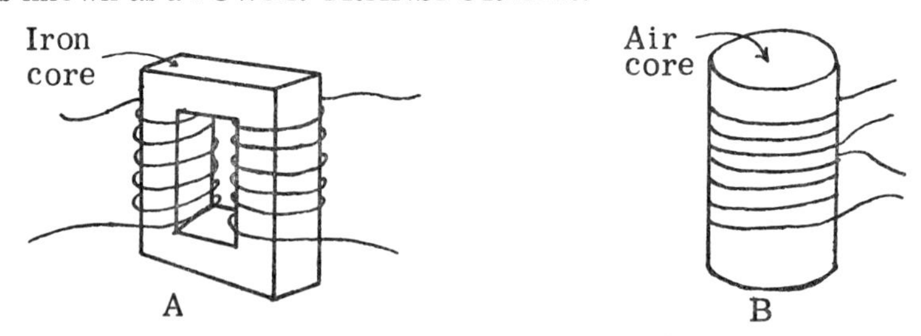

Figure 3-14. Iron-core and Air-core transformers.

In order for a power transformer to operate efficiently, the primary and secondary are wound on an iron core, as illustrated in Figure 3-14A.

Power transformers can only be used on AC because an alternating magnetic field is required to induce an EMF in the secondary. It is dangerous to apply DC to the power transformer primary. The primary has a low DC resistance and therefore, a high DC current will flow through it. This high current will either blow a line fuse or damage the transformer beyond repair.

3-14 IMPEDANCE

In a DC circuit, the only resistances that are present are those due to resistors or to the resistances inherent in long lengths of wire. The total resistance in a DC circuit is found by using the formulas on Pages 13 and 15.

In an AC circuit, there are additional resistances. In Paragraph 3-9, inductive reactance and capacitive reactance were discussed. These offer resistance to AC just as a resistor does. The total opposition by all "resistances" in an AC circuit is called IMPEDANCE. The unit of impedance is the OHM.

PRACTICE QUESTIONS — LESSON 3

1 Alternating current:
 a. flows only in one direction

b. reverses its direction of flow periodically
c. flows more in one direction than in the other
d. none of the above

2 The frequency of a sine wave is:
a. the time in seconds for one cycle
b. the amplitude of the wave
c. the number of cycles per second
d. the angle of rotation

3 One complete cycle has:
a. one positive alternation
b. one negative alternation
c. two negative alternations
d. a positive and negative alternation

4 One kilohertz equals:
a. $\frac{1}{1000}$ Hertz b. 25 Hertz c. 1000 Hertz d. 1 Hertz

5 A capacitor is:
a. two conducting plates connected by a wire
b. a resistance
c. two insulator plates separated by air
d. two conducting plates separated by a dielectric

6 A capacitor is:
a. a short-circuit for DC
b. an open-circuit for DC
c. offers little opposition to DC current flow
d. a DC generator

7 The unit of capacitance is:
a. henry b. coulomb c. farad d. gilbert

8 The unit of inductance is:
a. henry b. coulomb c. farad d. gilbert

9 If the two plates of a capacitor touch, the capacitor is said to be:
a. open b. good c. shorted d. a variable capacitor

10 In a power transformer, DC should not be applied to the primary because:
a. a counter EMF will be developed
b. no load will be present
c. a high DC current will flow
d. the efficiency will be poor

11 In a resonant circuit:
a. the inductive reactance is equal to the capacitive reactance.
b. the inductive reactance is less than the resistance.
c. the resonant frequency depends upon the inductance and resistance.
d. the harmonics are high.

SECTION II — LESSON 4
VACUUM TUBES AND SOLID STATE DIODES

4-1 THE DEVELOPMENT OF THE VACUUM TUBE

Edison's incandescent electric lamp was the forerunner of the modern electric bulb. It consisted of a resistance wire, called a filament, enclosed within a glass envelope. The air within the glass envelope had been removed to create a vacuum. The ends of the resistance wire protruded through the glass, as illustrated in Figure 4-1. If a current passes through the resistance wire, it will heat up and glow. We can then say that the filament wire has been heated to INCANDESCENCE. While working with his electric light, Edison discovered that the incandescent wire emitted, or boiled off, electrons. These electrons remained around the wire in the form of an electron cloud or SPACE CHARGE. This phenomenon of electron emission is known as the EDISON EFFECT, and is the basis of operation of all vacuum tubes.

4-2 ELECTRON EMISSION

Many metallic substances will emit electrons when heated to incandescence. In the previous paragraph, it was shown that the resistance wire in a light bulb emits electrons. These emitted electrons are wasted since they serve no useful purpose.

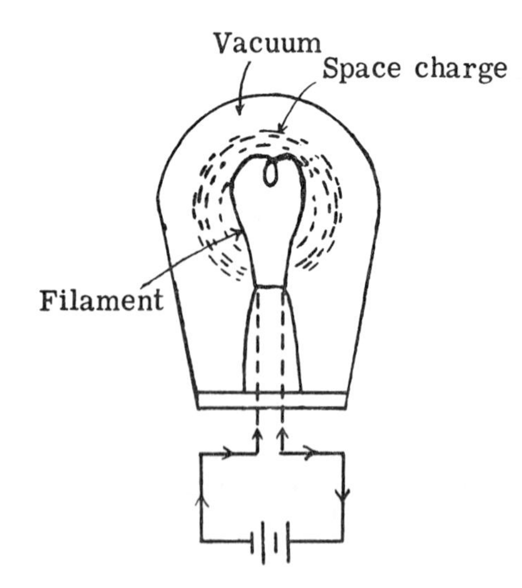

Figure 4-1. The electric lamp.

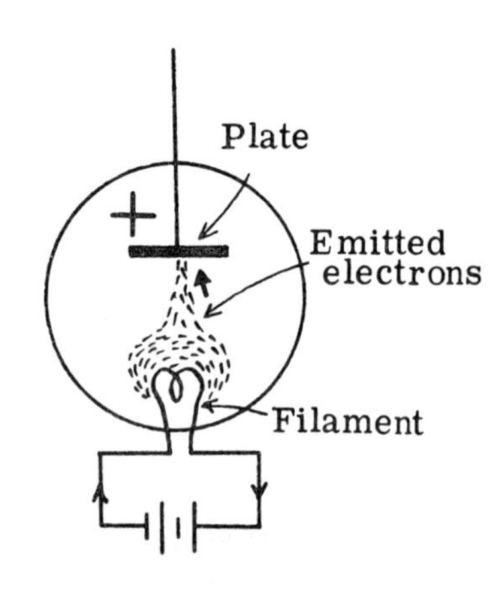

Figure 4-2. Positive plate attracting electrons.

The vacuum tube is similar to the light bulb in that it also contains a resistance wire which emits electrons when heated. The vacuum tube, however, is designed to make use of the emitted electrons. In addition to the resistance wire, the vacuum tube has a positively charged collector of electrons called THE PLATE. The positive plate attracts the emitted electrons, as illustrated in Figure 4-2.

The purpose of the battery in Figure 4-2 is to force current through the filament, thereby heating it.

4-3 THE CATHODE

The element in the vacuum tube which supplies the electrons for the tube operation is known as the CATHODE. The cathode, as does the resistance wire, emits or boils off electrons when energy in the form of heat is supplied to it. There are two different types of cathodes used in vacuum tubes. They are the directly heated and the indirectly heated types. We will now discuss these two types in detail.

1. THE DIRECTLY HEATED CATHODE. This type is also known by the name FILAMENT-CATHODE. An example of a filament-cathode is illustrated in Fig. 4-3. The heating current is passed directly through the cathode wire, which is made of tungsten. The current heats up the cathode wire, which then emits electrons from its surface. Directly heated filament-cathodes usually require very little heating power. They are, therefore, used in tubes designed for portable battery operation because it is necessary to impose as small a drain as possible on these batteries.

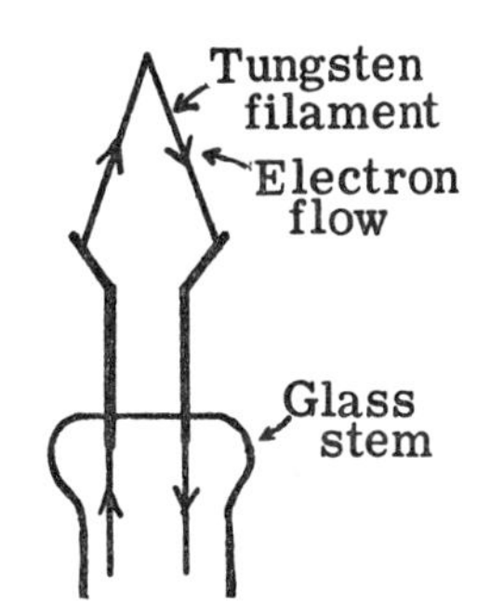

Figure 4-3. Directly heated cathode.

All vacuum tubes are classified by tube numbers. If you desire to know the purpose and characteristics of a particular tube, you simply look up its tube number in any tube manual for the information wanted.

2. THE INDIRECTLY-HEATED CATHODE. This type is also known as the HEATER-CATHODE and is illustrated in Figure 4-4A. Part A is a thin metal sleeve or cylinder coated with electron-emitting material. This sleeve is the cathode. Part B is a heater wire which is insulated from the sleeve. The heater is usually made of a tungsten material. Its sole purpose is to heat up the cathode sleeve to a high enough temperature so that the emitting material will boil off electrons. Note that the heater itself does not give off the electrons. The heater wire is known as the filament. Figure 4-4B shows the schematic symbol for the heater-cathode.

Almost all present day receiving tubes designed for AC operation are of the indirectly-heated cathode type. We will always refer to the electron emitting surface as the CATHODE and the heater as the FILAMENT.

4-4 FILAMENT OPERATING VOLTAGE

The first number in a tube designation usually indicates the proper filament operating voltage. For example, a 6H6 tube should have its filament operated at 6.3 volts. All filaments should be operated at their designated operating voltages, which are determined by the manufacturer. If the filament is operated above its rated voltage, the excessive current will shorten the filament life. Operating the filament below its rated voltage will decrease electron emission and lower the tube operating efficiency.

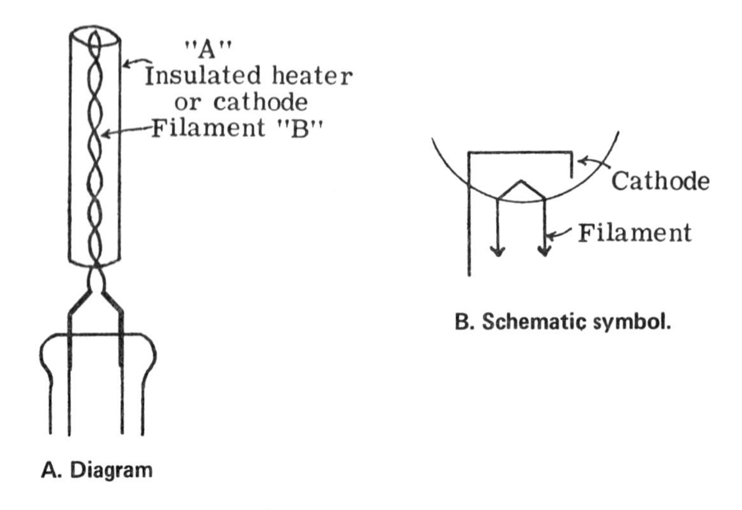

A. Diagram

B. Schematic symbol.

Figure 4-4. The indirectly heated cathode.

4-5 THE DIODE

Let us see how electrons emitted from the cathode can be collected and made to do useful work. Electrons are negatively charged and will be attracted by a positively charged object. Therefore, it a positively charged object called a PLATE is put into the vacuum tube, it will serve as a collector of

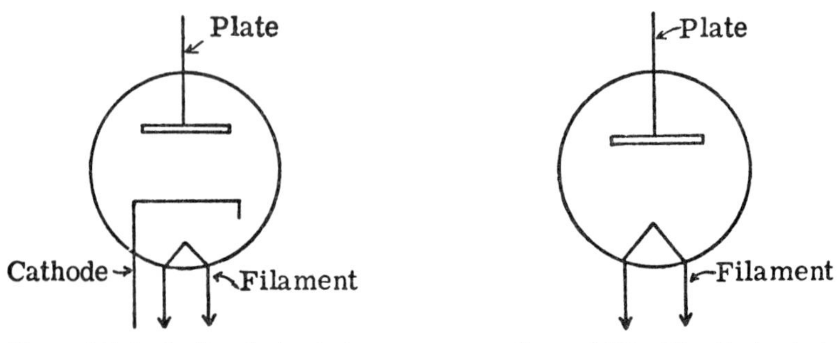

Figure 4-5 A. Indirectly heated.

Figure 4-5B. Directly heated.

electrons. A vacuum tube which contains a plate and a cathode is known as a DIODE. The schematic symbol for the diode is shown in Figure 4-5. B is a directly-heated diode and A is an indirectly-heated diode.

The plate and the cathode are known as the ELEMENTS of the vacuum tube. The diode is, therefore, a two-element tube. The heater of the indirectly-heated tube is not counted as a separate element.

4-6 THE DIODE AS A CONDUCTOR

Figure 4-6 illustrates a simplified schematic of a diode with the plate connected to the positive terminal of a battery; the cathode is connected through a switch to the negative terminal. The instant the switch is closed, the ammeter in the circuit will register a current flow, indicating that electrons are flowing from the cathode to the plate. The diode is said to be CONDUCTING. The diode conducts because the plate is positive with respect to the cathode. Therefore, the plate attracts to it the negatively charged electrons emitted by the cathode. The electrons flow from the plate to the positive terminal of the battery. They then flow through the battery and back to the cathode, where they once more can be emitted to the plate. If the battery voltage is increased, the plate will become more positive and will, therefore, attract more electrons. Consequently, the ammeter will register a larger current flow. Conversely, if the battery voltage is decreased, the plate will attract less electrons and the ammeter will register a smaller current flow.

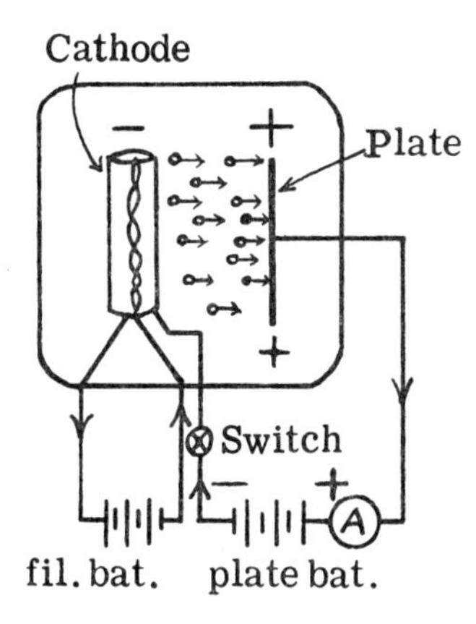

Figure 4-6. Electron flow when plate is positive.

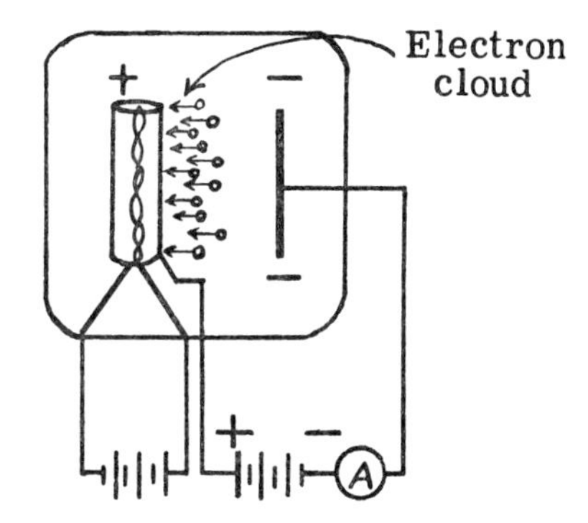

Figure 4-7. Diode action when plate is negative.

When the diode conducts, it presents a very low resistance path between the cathode and plate. For all practical purposes, we can consider a conducting diode as a closed switch (short circuit) between the cathode and plate.

4-7 THE DIODE AS A NON-CONDUCTOR

If we reverse the battery connections, as shown in Figure 4-7, the plate becomes negative and the cathode positive. Since the negative plate will not attract electrons, the diode will NOT CONDUCT. The diode, therefore, acts like an open switch (open-circuit) and permits no current flow. The ammeter will consequently read zero amperes. The emitted negatively-charged electrons are repelled by the negative plate and remain close to the cathode where they form an ELECTRON CLOUD. The cloud of electrons around the cathode is known as a SPACE CHARGE. The space charge, by virtue of its large negative charge, prevents the plate from receiving any more electrons. When the plate becomes positive once again, the space charge is rapidly dispelled since it is attracted to the plate. The cathode is free once again to emit electrons to the plate.

Let us now summarize the operation of the diode:

1) electrons flow in one direction only - from cathode to plate.

2) electron flow to the plate will take place only when the plate is positive with respect to the cathode.

3) the current flow will vary with the plate to cathode voltage.

4) the diode acts as a conductor (short circuit) when the plate is positive.

5) the diode acts as a non-conductor (open circuit) when the plate is negative.

4-8 SEMICONDUCTOR DIODES

In the early days of radio, a receiver known as a "crystal set" was very popular. The original crystal sets used earphones because the amplifier tubes that were required to operate a speaker were quite expensive.

When tubes became less costly and demand for speaker-operated receivers became strong, the crystal set began to disappear. However, the "crystal" itself, in a somewhat different form, has made a strong comeback. Today, it is known as a solid-state semiconductor.

The diode semiconductor is used in just about every radio and television set. Its popularity is based on the fact that, unlike the tube, it does not require a socket; it is light in weight and can be soldered directly into a circuit. Furthermore, it has no filament and so its operation is instantaneous. The semiconductor diode works both as a rectifier and a detector. Let us see how this type of diode functions, why it works and also, how it differs from a vacuum tube.

4-9 INSULATORS

Certain substances and certain elements are good conductors; others are good insulators. Copper wire is a good conductor, while glass is a good insulator. The element iron is put in the conductor class, while other elements, such as pure germanium, selenium or pure silicon, are insulators.

We can modify the characteristic of any element by mixing in other elements. For example, we can add boron, antimony or arsenic to pure germanium or pure silicon and change these elements from non-conductors to conductors.

4-10 DOPING

The addition of antimony to either germanium or silicon, is known as doping. Since the germanium or silicon, at the start, is as pure as it can be made, the added element is referred to as an impurity. It takes a very small amount of impurity to modify the germanium or silicon so that they are no longer insulators.

All elements are made up of atoms and each atom contains a nucleus surrounded by one or more rings of electrons. By diffusing certain elements, such as antimony, into pure germanium or silicon, we change the total number of electrons in the germanium or silicon. These "doped" elements now have more electrons than they originally had. Since electrons are negatively charged, we refer to them as negative germanium or negative silicon. We abbreviate negative germanium as n-germanium or n-type germanium. The antimony, diffused into the pure germanium or silicon is referred to as a "doner" element since it has, in effect, donated or contributed electrons.

We can diffuse substances that have a deficiency of electrons into germanium or silicon. An element, such as boron, for example, likes to borrow electrons and so, when mixed with germanium or silicon, will take electrons from these substances. For this reason, we call boron an acceptor element - it accepts or takes away electrons.

When a substance or an element loses electrons, it is no longer neutral. It has become less negative or, stated in another way, has become more positive. We can also refer to an atom that has lost an electron as a "hole". The "hole" is the positive charge around the atom due to the removal of one of its electrons by the acceptor. Germanium or silicon, doped with boron, is referred to as positive germanium or positive silicon. We abbreviate this as p-germanium or p-silicon or p-type. We can represent n-type or p-type pictorially, as shown in Figure 4-8. The minus signs indicate an excess of negative charges; the plus signs tell us that we have a shortage of electrons or an excess of positive "holes".

Figure 4-8. N-type and P-type germanium or silicon.

N-type material does have a small amount of positive holes along with the vast number of electrons. We refer to these positive holes as "minority carriers". The electrons are referred to as "majority carriers". A similar situation exists in P-type material. There are a few negative (electron) minority carriers, along with a large number of positive (holes) majority carriers.

4-11 THE SEMICONDUCTOR DIODE

If we take a block of p-type germanium and a block of n-type germanium (or silicon) and put them together, we will have a semiconductor diode. The semiconductor diode is referred to as a solid-state device.

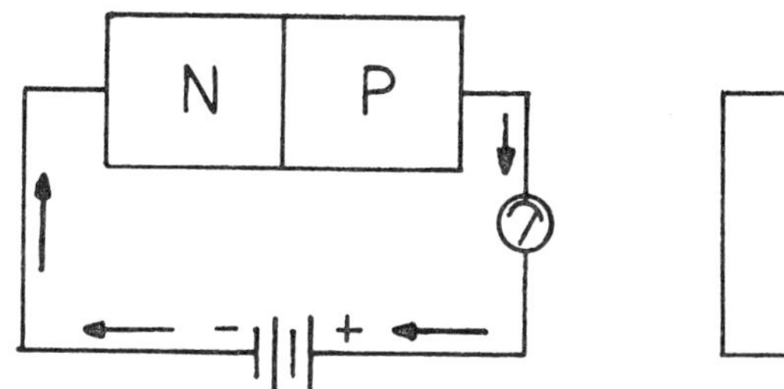

Fig. 4-9A. Forward biasing.

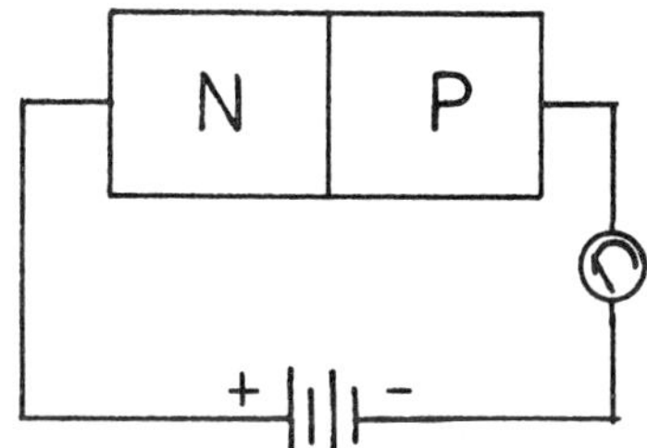

Fig. 4-9B. Reverse biasing.

In Figure 4-9, we have a battery connected across our two blocks of doped germanium or silicon. In drawing A, the negative terminal of the battery is connected to the n-type material, while the positive terminal is connected to the p-type material. Connecting a voltage in this manner is known as biasing. Electrons will now flow from the battery, through the n-type material, into the p-type material and back to the battery. The reason for the current flow is that the battery urges or forces electrons into the n-type block, which already has more electrons than normal. The electrons migrate over to the p-type block since this region is more positive and attracts them. However, as electrons leave the p-type block to the battery, more electrons from the n-type block cross the juntion between the two blocks, and so the process is a continuous one. The current that flows is referred to as a forward current. The voltage producing this current is then called a forward voltage or forward bias.

Now examine drawing B. The only difference is that we have reversed the polarity of the battery. As a result, very little current flows. The small current that does flow, moves in an opposite direction to the way it previously moved. We, therefore, call it a reverse current and the battery voltage is referred to as a reverse voltage or reverse bias.

Aside from the fact that the semiconductor diode in Figure 4-9 does have a small amount of reverse current, its basic action is very much like the vacuum tube diode described earlier in this lesson. Note that there is no filament or cathode to be heated and so, unlike the tube, the semiconductor diode does not get warm or hot when operating. Therefore, since we do not need to wait for the filament or cathode to get hot enough to emit electrons, the semiconductor diode acts at once.

Figure 4-10A illustrates the symbol for the standard type of diode. Figure 4-10B is the symbol for a special type of diode called a Zener diode.

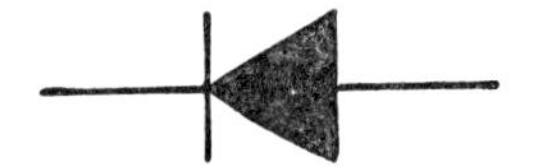

Fig. 4-10A. Common diode.

Fig. 4-10B. Zener diode.

PRACTICE QUESTIONS — LESSON 4

1 The diode tube has:
   a. one element   b. two elements   c. three elements   d. four elements

2 Electron emission:
   a. is undesirable in vacuum tubes
   b. is necessary for the operation of a light bulb
   c. can only take place when the filament is cold
   d. is the giving off of electrons by a metal when heated to incandescence.

3 The plate is:
   a. a positively charged collector of electrons
   b. a positively charged emitter of electrons
   c. not necessary for the operation of a diode
   d. connected directly to the cathode

4 The cathode:
   a. is not necessary for the operation of a diode
   b. is a positively charged collector of electrons
   c. repels electrons
   d. emits electrons for tube operation

5 The diode acts as an open-circuit:
   a. when the tube conducts
   b. when the plate is negative with respect to the cathode

c. when the plate emits electrons
d. when the cathode is negative with respect to the plate

6 A diode tube allows current to flow:
a. only from cathode to plate
b. only from cathode to heaters
c. in either direction
d. straight up

7 When the plate of a diode is positive, relative to the cathode:
a. current will flow from plate to cathode
b. the cathode stops emitting
c. the tube conducts
d. an electron cloud forms

8 What are the majority carriers in N-type material?
a. holes
b. electrons
c. protons
d. neutrons

9 Which of the following creates N-type material when diffused into germanium?
a. antimony
b. boron
c. selenium
d. copper oxide

10 In semiconductor conduction, a "flow of holes" refers to:
a. a flow of positive carriers
b. a flow of electrons
c. a flow of protons
d. a flow of neutrons

SECTION II – LESSON 5
RECTIFICATION, FILTERING

5-1 RECTIFICATION

Vacuum tubes in receivers and transmitters will only operate when connected to a direct current source of power. Portable radios, for example, are energized by batteries which are in themselves a source of direct current. As previously noted, the electrical power that is delivered to most homes throughout the country today is alternating current. If we were to connect the tubes in our radios directly to the AC wall outlet, the radio would not operate because a radio tube needs a source of DC power. We all know that our radios DO operate when we plug them into the AC socket. Obviously, there must be something in the radio which converts the alternating current into direct current. The device in a radio which converts the alternating current into direct current is known as a RECTIFIER. The process of conversion is known as RECTIFICATION.

5-2 THE DIODE AS A HALF-WAVE RECTIFIER

The ability of the diode to pass current in only one direction makes it possible to convert alternating current into direct current. Let us see how this takes place. Figure 5-1 illustrates a simple diode rectifier circuit.

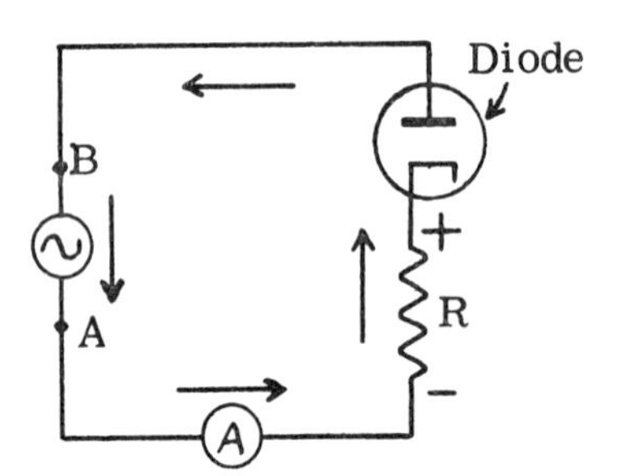

Figure 5-1. Diode used as half-wave rectifier.

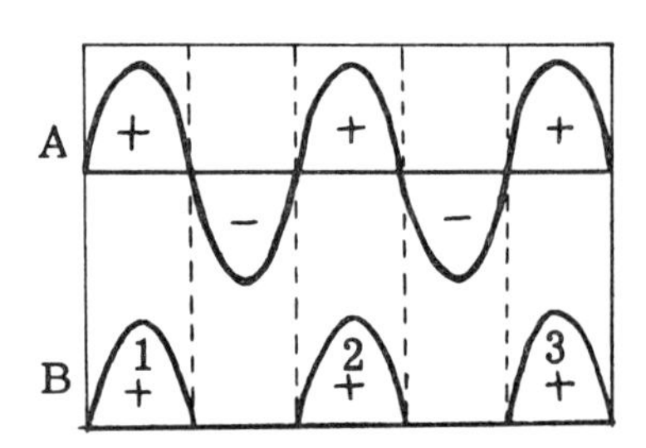

Figure 5-2. Half-wave rectifier wave-forms.

When terminal "B" of the AC generator is positive with respect to terminal "A", the diode plate becomes positive with respect to its cathode. The diode, therefore, conducts current in the direction indicated by the arrow. The DC milliammeter will deflect to the value of the current flow.

On the next half of the alternating current cycle, the polarity of the generator will be reversed, making the plate negative with respect to the cathode. The diode becomes a non-conductor, and the current will stop flowing. On the next cycle, the polarity of "A" and "B" will again reverse itself. The diode will conduct and once again, current will flow. Examination of Figure 5-2 shows what is really happening. Figure 5-2A is the sine wave which is generated across the terminals of the AC generator. Figure 5-2B is the wave which is obtained across the load resistor R. Alternations 1, 2 and 3 are the positive halves of the cycle when the plate of the diode is positive with respect to the cathode. At that time, the diode conducts and acts as a short circuit. The positive voltage alternations are therefore impressed directly across the resistor R. During the negative half of each AC cycle, the tube does not conduct and is an open circuit. During these times, there is no voltage developed across the resistor since there is no current flow. The current through the resistor is therefore a pulsating direct current, and the voltage across the resistor is a pulsating direct voltage. Even though the current flows in spurts or pulses through the resistor, the current is DC because it flows only in ONE direction. This action of the diode in passing only one-half of the AC input wave to the load resistor is known as HALF-WAVE RECTIFICATION.

The ends of the load resistance have been marked with a polarity because electrons are entering and emerging from this resistance. The end they enter becomes more negative than the end from which they emerge. The pulsating direct-voltage, if properly filtered, can be utilized to operate a radio receiver.

A transformer can be considered as an AC generator. We can, therefore, replace the AC generator of Figure 5-1 with a transformer as shown in Figure 5-3, without altering the operation of the circuit.

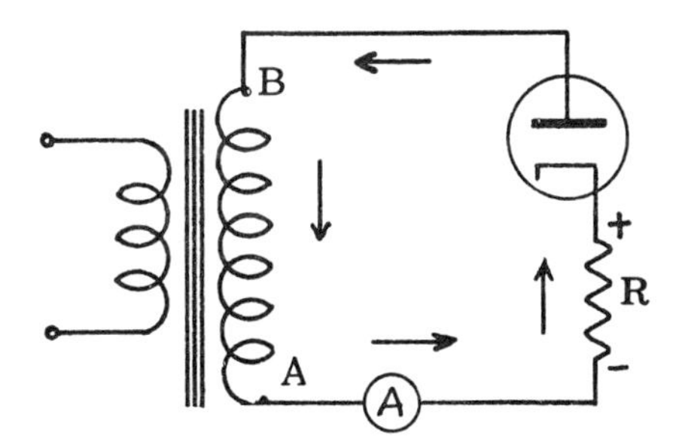

Figure 5-3. Diode used as half-wave rectifier.

5-3 THE DIODE AS A FULL-WAVE RECTIFIER

In half-wave rectification, only the positive half of the AC input is used. The negative alternations are completely cut off and wasted. If we could somehow utilize the negative as well as the positive alternation, we would be operating our rectifying system more efficiently. This is accomplished in full-wave rectification.

We can modify the half-wave rectifier circuit of Figure 5-3 by adding another diode and center-tapping the transformer secondary. The resulting circuit is illustrated in Figure 5-4. The cathodes of the diodes are connected together, and the circuit is known as a FULL-WAVE RECTIFIER.

The operation of a full-wave rectifier is as follows: When an AC voltage is impressed across the primary of the transformer, an AC voltage will be induced across the secondary. When point "A" is positive with respect to point "B", the plate of diode #1 is positive and the tube conducts. The electrons flow through the transformer, from A to C, out of C into the load resistance R_L, and back to the cathode of diode #1. During all this time, the plate of diode #2 is negative and does not conduct. On the next half of the AC cycle, the bottom of the transformer, point "B", goes positive while the top, point "A", goes negative. The plate of diode #2 is now positive and the plate of diode #1 is negative. Now, diode #2 conducts and diode #1 does not. The

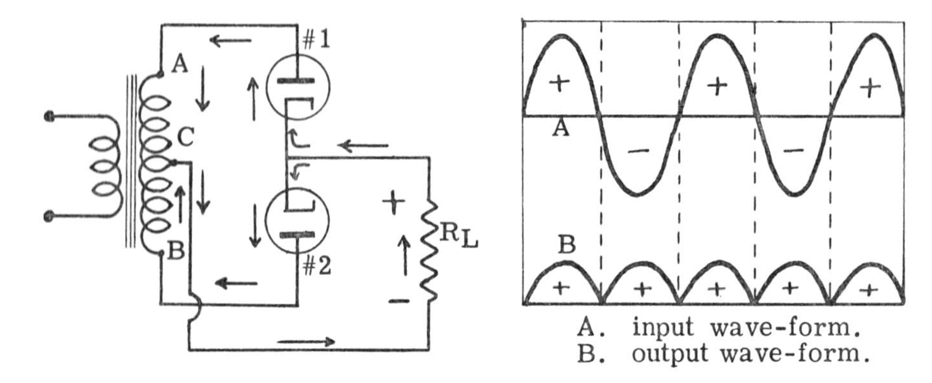

Figure 5-4. Full-wave rectifier.

Figure 5-5. Full-wave rectifier wave-forms.

electrons flow through the transformer from B to C, into the load resistor R_L, and back to the cathode of diode # 2. Notice that the current flows through the resistor in the same direction during both the positive and negative halves of the input cycle. We have very definitely used both halves of the AC input cycle, and have accomplished full-wave rectification. Figure 5-5A shows the AC across the transformer secondary. Figure 5-5B shows the pulsating DC flowing through the load. Compare this output with the rectified wave picture of Figure 5-2B.

The pulsating DC output of Figures 5-2 and 5-5 are not pure DC. The amplitude of a pure DC wave is constant. In order to change pulsating DC into pure DC, a filter is used. The filter, which contains capacitors and chokes, removes the ripple from the pulsating DC and gives us pure DC. The output of a full-wave rectifier is easier to filter than that of a half-wave rectifier.

5-4 SUMMARY OF RECTIFICATION

(1) A single diode may be used as a half-wave rectifier for converting AC to DC. Only half of the input AC wave is used, and the full voltage of the secondary of the power transformer is obtained as useful DC output.

(2) A double diode may be used as a full-wave rectifier. Both halves of the AC wave are used, and the output voltage is only half of the total transformer secondary voltage.

PRACTICE QUESTIONS — LESSON 5

1 A rectifier is used to:
  a. change DC to AC
  b. change AC to DC
  c. increase the ripple frequency
  d. improve voltage regulation

2 The primary characteristic of a diode is:
  a. it passes current in both directions
  b. it passes current in one direction only
  c. it does not pass current
  d. it is used to convert DC to AC

3 A half-wave rectifier contains:

a. one diode b. two diodes c. no diodes d. an insulator

4 A full-wave rectifier requires a minimum of:
a. one diode b. two diodes c. three diodes d. four diodes

5 A diode becomes a non-conductor when:
a. the plate is positive
b. the plate is negative
c. the load resistor is connected
d. the current source is on

6 The output voltage of a full-wave rectifier is equal to approximately:
a. the voltage of the full transformer secondary.
b. the voltage of half the transformer secondary.
c. the voltage of the full transformer primary.
d. the voltage of half the transformer primary.

SECTION II — LESSON 6
TUBES AND TRANSISTORS

6-1 INTRODUCTION

Lessons 4 and 5 covered the construction and purpose of a diode vacuum tube. We studied the action of the diode vacuum tube as a rectifier in changing alternating current to direct current. We will now go into the details of the operation of the vacuum tube when used as an amplifier. An amplifier makes larger, or amplifies, small AC voltages. The vacuum tubes that are used for amplification purposes are three, four and five element tubes. The three element tube is called a TRIODE; four and five element tubes are called TETRODES and PENTODES respectively. We shall now proceed to study each one of these tubes in detail.

6-2 THE TRIODE

The TRIODE is different from the diode in that it contains one more element. This new element is called the CONTROL GRID. The control grid is a thin piece of wire wound in the form of a spiral mesh which surrounds the cathode. Electrons emitted by the cathode can pass easily through the grid structure and onto the plate. Figure 6-1A shows the actual physical arrangement of the cathode, grid and plate structure in a typical triode. Notice that the grid is placed much closer to the cathode than to the plate.

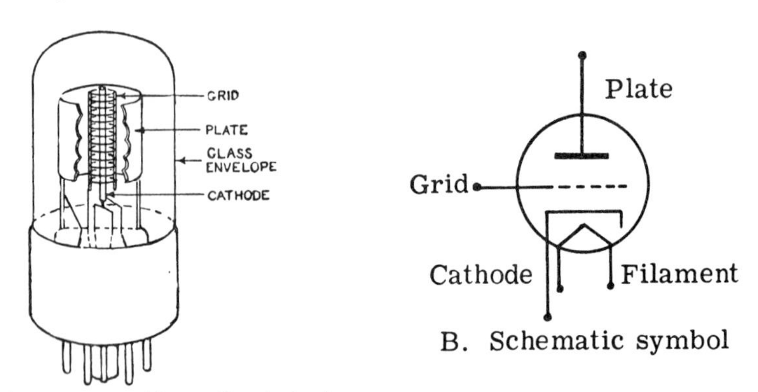

Figure 6-1. The triode.

Figure 6-1B illustrates the schematic representation of the triode. The grid is shown by means of a dotted line between the cathode and plate.

6-3 OPERATION OF A TRIODE

Figure 6-2 shows a triode circuit which is used to study the effect of grid voltage variations upon the plate current. The symbol for the plate voltage is E_p. The plate voltage is measured between the plate and cathode. The symbol for plate current is I_p. Plate current is measured by placing an ammeter in series with the plate circuit. E_g is the symbol for control grid voltage, measured between the control grid and the cathode. All tube voltage measurements are taken with the cathode as a reference point.

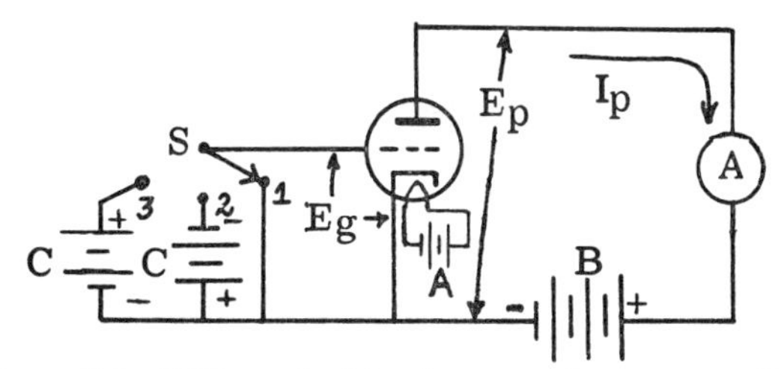

Figure 6-2. Efffect of grid voltage on plate current flow.

Note the letters A, B and C that are near the battery symbols in the diagram (Figure 6-2). These letters indicate the voltages applied to the different elements in the tube. The "A" voltage is applied to the heater or filament. The "B" voltage is applied to the plate; the "C" voltage is applied to the grid. "S" is a three-position switch in the control grid circuit. With the switch in position #1, the control grid is connected directly to the cathode. With the switch in position #2, the control grid is connected to the negative terminal of the battery. With the switch in position #3, the control grid is connected to the positive terminal of a battery. Let us see how changes in the control grid voltage affect the operation of the triode. With the switch in position #1 and the plate positive, electrons will flow from the cathode through the grid structure to the plate. Since the grid is connected directly to the cathode, it will not affect the flow of plate current. Therefore, all the emitted electrons will pass through the grid and onto the plate.

If the switch is thrown to position #2, the grid becomes negative with respect to the cathode. The negatively charged grid will repel many of the

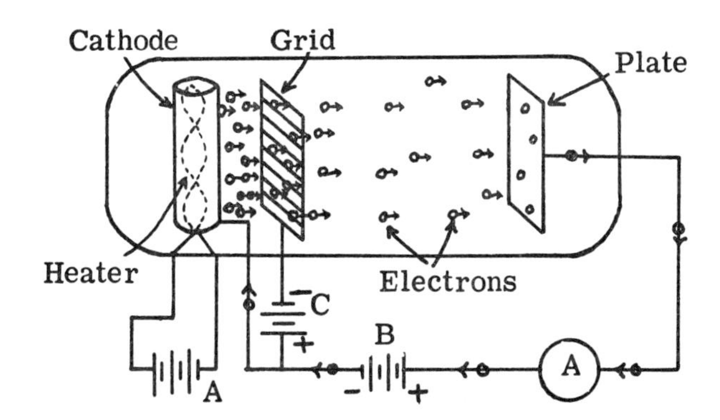

Figure 6-3. Effect of negative grid on plate current flow.

negatively charged electrons back into the area surrounding the cathode. Hence, the number of electrons which are able to reach the plate is reduced. This effect is illustrated in Figure 6-3. The milliammeter in the plate circuit will show a reduction in plate current when the grid voltage is changed from a zero voltage to a negative voltage.

If the switch is thrown to position #3, the grid becomes positive with respect to the cathode. The plate current will increase since the positive control grid attracts the negative electrons and allows many more electrons to be drawn to the plate than it did in switch position #1 and #2. A positive grid actually pulls electrons from the cathode to the plate. Thus, we see how the control grid acts as a control valve for plate current flow. As we vary the voltage on the grid, the plate current varies. The control grid, therefore, controls the flow of electrons to the plate.

6-4 THE TRIODE AS AN AMPLIFIER

In Paragraph 6-1, it was stated that multi-element tubes are used to amplify weak signals. We will now proceed to study the exact manner in which a triode tube amplifies a signal voltage that is applied to its control grid.

The control grid is physically much closer to the cathode than the plate is. The grid voltage will, therefore, have a greater effect on the plate current than will the plate voltage. A small change in grid voltage will cause a large change in plate current; whereas, a small change in plate voltage causes a small change in current. Let us see, graphically, how a changing voltage such as an AC signal on the grid of a triode, causes the plate current to vary. Figure 6-4 illustrates a triode whose plate is connected to a fixed B+ voltage. The grid is in series with an AC generator and a fixed bias voltage. The total voltage between the grid and cathode will always be the sum of the signal voltage and the bias voltage.

In order to understand how a tube amplifies, we must refer to the grid voltage, plate current curve of Figure 6-5. It is not necessary to understand thoroughly the derivation of the curve; however, the reader should know that the curve is simply a graph that tells us what the plate current is when a certain voltage is applied to the grid. In the discussion, we will assume a signal voltage of one volt.

Let us see what happens on the positive half-cycle of the AC signal. Since the signal voltage of +1 volt and the -3 volts of bias are in series, the resultant voltage between grid and cathode will be -2 volts. (The sum of +1 and -3 = - 2). On the negative half of the AC cycle, - 4 volts will be applied between the grid and cathode of the tube. (The sum of -1 volt and -3 volts = -4 volts). From the Ip - Eg curve of Figure 6-5, it can be seen that when there is no AC signal applied to the grid, the plate current will be fixed at 8 milliamperes because of the three volts of bias supplied by the bias battery. The value of 8 milliamperes is obtained from the curve by working vertically from the -3 volts point on the grid voltage line until the curve is reached. From this point we go straight across until we hit the vertical plate current line. In this case, we reach the vertical line at 8 milliamperes. On the peak of the positive half of the AC signal (when there are -2 volts on the grid), the plate current rises to 10 milliamperes. On the negative half of the incoming signal (when there are -4 volts on the grid), the plate current decreases to 6 milliamperes. Note that the waveform of the plate current variation is an exact reproduction of the AC signal applied to the grid of the tube. A 2 milliampere variation is caused in the plate current by applying a one volt signal to the grid.

Thus far, we have converted grid voltage variations into plate current variations. In order to make use of these plate current variations, some device must be placed in the plate circuit to act as a load across which the varying plate current will develop a varying voltage. The plate load may be a

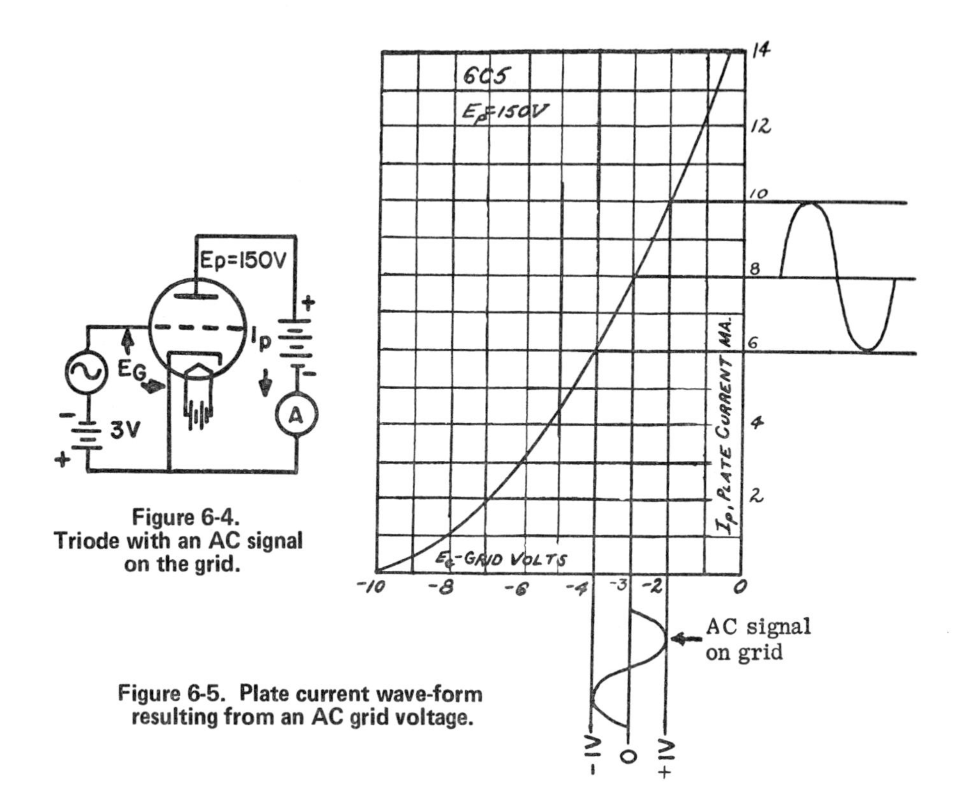

Figure 6-4.
Triode with an AC signal on the grid.

Figure 6-5. Plate current wave-form resulting from an AC grid voltage.

resistor, an inductor or a tuned circuit. Figure 6-6 shows a resistor used as a plate load in a triode amplifier circuit. Except for the plate load resistor, this circuit is the same as that in Figure 6-4. As we explained before, the 1 volt signal will cause a total plate current variation of 4 milliamperes (from 6 to 10 ma.). This 4 ma. variation will cause a total voltage variation of 40 volts to be produced across the 10,000 ohm resistor. This can easily be proven by

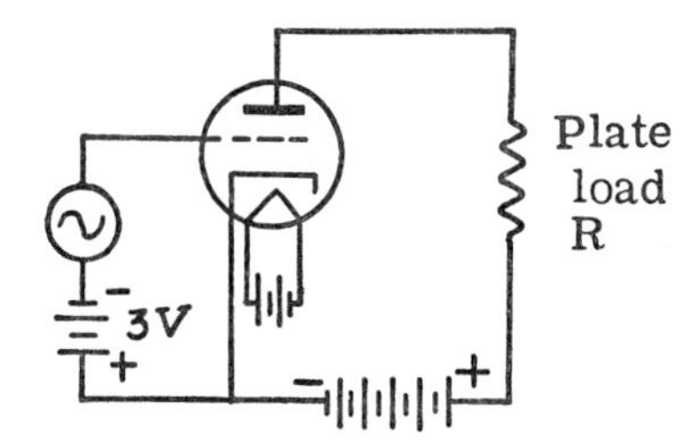

Figure 6-6. Triode using a resistor as a plate load.

Ohm's law. One form of Ohm's law states that: E = I x R, E = .004 x 10,000 E = 40 V. Thus it can be seen that a 2 volt AC signal (2 volts peak to peak) can produce a 40 volt variation in the plate circuit. In other words, the original signal or variation that was applied to the grid has been AMPLIFIED twenty times.

$$\left(\frac{40}{2} = 20\right)$$

From Figure 6-5 it can be seen that the voltage variation in the plate circuit is not only an amplified, but also a faithful reproduction of the grid signal. The circuit in Figure 6-6 is, therefore, the basis for all amplification circuits in radio and television.

Figures 6-7 and 6-8 illustrate schematic diagrams of the tetrode and pentode vacuum tubes. Their operation is similar to that of the triode. The extra elements make up for some of the deficiencies of the triode.

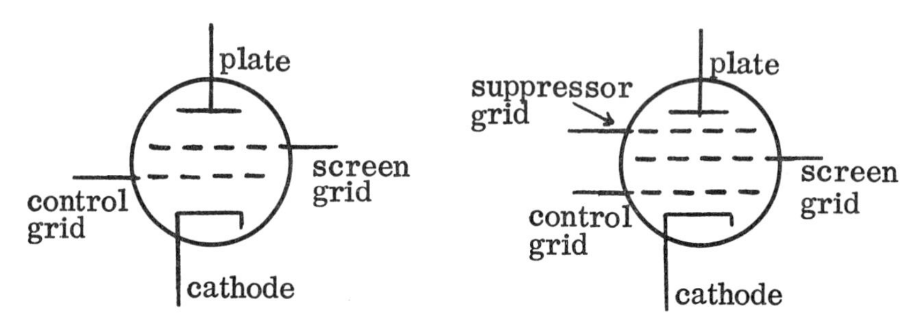

Fig. 6-7. Tetrode tube.

Fig. 6-8. Pentode tube

6-5 PLATE POWER INPUT

The DC plate power input to a tube is equal to the product of the DC plate voltage and the DC plate current. For instance, if the plate voltage is 750 volts and the plate current is 150 milliamperes, the power input is 112.5 watts. We arrive at this in the following manner:

$$\text{Power input in watts} = E_p \times I_p = 750 \times .15 = 112.5 \text{ watts}$$

Note that the 150 ma. was changed to amperes by moving the decimal three places to the left.

6-6 THE TRANSISTOR

From what we have learned about tubes in the previous lesson, we know that we can make a diode into a triode by adding a new element called the control grid. This single, added electrode makes a tremendous difference, for the triode tube can amplify, whereas the diode cannot. Thus, it was the triode that advanced receivers from the headphone state to speaker operation.

In a similar manner, semiconductor triodes that amplify can be made from the semiconductor diodes previously described. Figure 6-9 illustrates a pair of P-N diodes placed back to back. The drawing shows that the two diodes have been pushed together.

Combining the two diodes of Figure 6-9 gives us the unit shown in Figure 6-10. We still have our pair of semiconductor diodes because we can divide the combined center of N section into two parts. Figure 6-10 is known as a transistor. The transistor of Figure 6-10 is called a P-N-P type. However, if we go back to our two diodes of Figure 6-9 and turn them around, we can have the two N-sections on the outside and the two P-sections joining each other. In this case, the transistor would be an N-P-N type. Figure 6-11 shows the N-P-N type.

Combining the N and P material in the above manner gives us the basic transistor. Transistors are capable of amplifying. Figure 6-12A shows an elementary transistor amplifier circuit. The E stands for emitter, the B for base

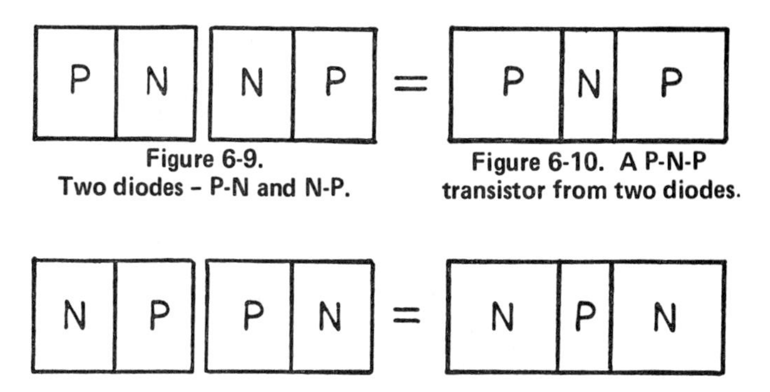

Figure 6-9.
Two diodes - P-N and N-P.

Figure 6-10. A P-N-P transistor from two diodes.

Figure 6-11. An N-P-N transistor from two diodes.

and C for collector. The input signal is applied between the emitter and base; the amplified output signal is taken from the load resistor that is across the collector and base.

6-7 TRANSISTOR SYMBOLS

There are two transistor symbols; one for the P-N-P type and the other for N-P-N. The symbol in the circuit of Figure 6-12A is that of an N-P-N transistor. The P-N-P symbol is exactly the same, except that the arrow in the emitter points inward, as shown in Figure 6-12B.

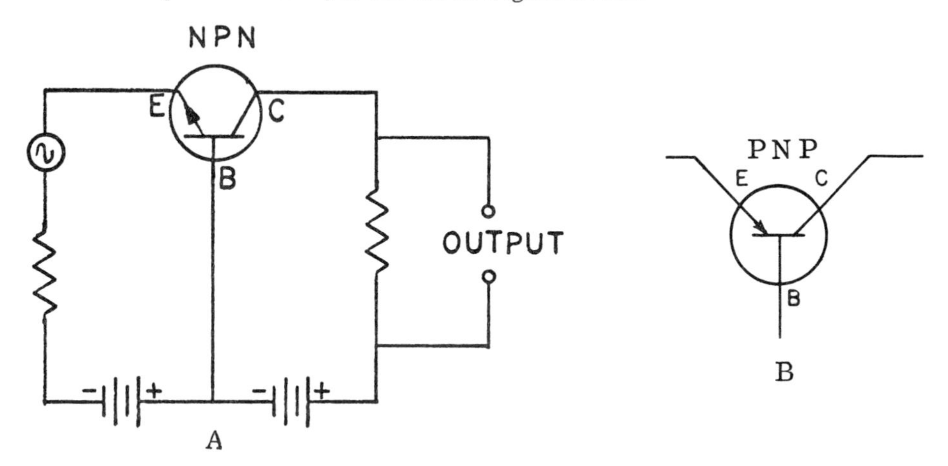

Figure 6-12. A basic transistor amplifier circuit.

PRACTICE QUESTIONS — LESSON 6

1 Increasing the negative voltage on the control grid will:
   a. decrease the plate voltage
   b. increase the plate current
   c. decrease the plate current
   d. have no effect on the plate current

2 The grid voltage of an indirectly heated tube is the voltage between the:
   a. grid and cathode
   b. grid and plate
   c. grid and filament
   d. grid and B+

3 The tube that cannot amplify is the:

a. pentode b. tetrode c. triode d. diode

4 An increase in positive grid voltage causes the plate:
a. current to decrease
b. current to increase
c. resistance to increase
d. voltage to increase

5 The DC plate power input to a tube having a plate voltage of 800 volts and a plate current of 85 ma. is:
a. 6,800 W. b. 68,000 W. c. 680 W. d. 68W.

6 The DC plate power input to a tube having a plate voltage of 550 volts and a plate current of 120 ma. is:
a. 66,000 W. b. 670 W. c. 66 W. d. 660 W.

7 A triode has:
a. no grids b. one grid c. two grids d. three grids

8 A pentode has:
a. no grids b. one grid c. two grids d. three grids

9 A transistor's base is similar to a tube's:
a. filament. b. collector. c. grid. d. cathode.

10 Which of the following are found in a transistor?
a. plate and emitter
b. collector and cathode.
c. emitter and diode.
d. collector and emitter.

SECTION II – LESSON 7
AUDIO AMPLIFIERS

7-1 INTRODUCTION

At this point, we understand that when a small amplitude signal is applied to the grid of a triode or pentode, it will be amplified and will appear many times larger in the plate circuit. This property of grid-controlled vacuum tubes makes possible their use as AMPLIFIERS. An amplifier may be defined as a device which transforms a small input signal into a large output signal.

7-2 AMPLIFIER APPLICATION

Amplifiers find many practical applications. For example, the signal that is developed in the crystal pickup of a record player is much too weak to be applied directly to a loud-speaker. This weak signal must first be amplified (made larger) before it can properly drive a loud-speaker. A lecturer addressing an audience in a large auditorium must have his voice amplified in order for him to be heard by everyone in the hall. The amplifier that accomplishes this is called a PUBLIC ADDRESS SYSTEM. Amplifiers are also extensively used in fields such as motion pictures, electrical recording and photoelectronics. Since amplifiers find such a wide application, it is important that we thoroughly understand their operation.

7-3 A TYPICAL AMPLIFIER

Figure 7-1 illustrates a simple amplifier. This amplifier consists of the following basic components:

1. a vacuum tube, such as a triode or pentode.
2. a power source for the filament of the vacuum tube, which is called an "A" supply.
3. a source of DC power (B+) for the plate circuit of the vacuum tube, which is called a "B" supply.
4. a bias voltage supply called a "C" supply.
5. a means of coupling the amplified signal from the plate circuit to the load. In Figure 7-1, the transformer couples the signal from the plate to the speaker.

When an amplifier consists of one tube, it is called a one-stage amplifier. If additional amplification of the signal is required, a second vacuum tube is connected in series with the first tube. The amplifier is then a two-stage amplifier; the vacuum tubes are said to be connected in CASCADE.

7-4 AMPLIFIERS USED IN RADIO RECEIVERS

The modern radio receiver uses two types of amplifiers in its operation. They are:

1. THE RADIO-FREQUENCY (RF) AMPLIFIER: This amplifier amplifies the weak radio-frequency signals picked up by the antenna of the receiver. A radio frequency signal is a high frequency radio wave (usually above 30 kiloHertz (kHz) which is sent out into space by the radio transmitter.

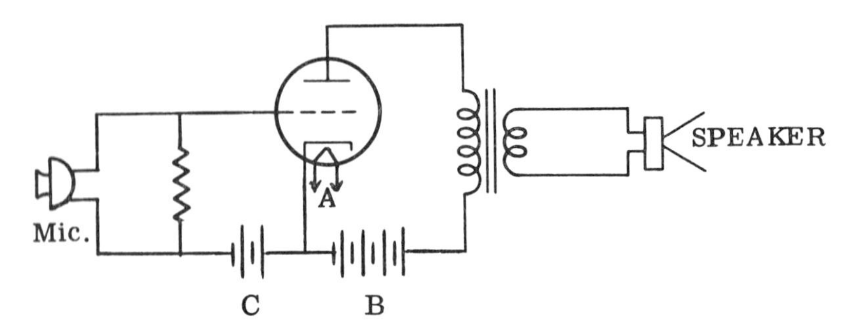

Fig. 7-1. Simple one-stage amplifier circuit.

2. THE AUDIO-FREQUENCY (AF) AMPLIFIER: This amplifier amplifies the sound frequencies or audio frequencies which are then applied to the loud-speaker. Audio frequencies are in the range between 16 and 16, 000 Hz.

7-5 SOUND

An audio amplifier is used to amplify the small signal output of a microphone. The action of a microphone depends upon certain characteristics of a sound wave. We have, therefore, reached a point in our discussion of amplifiers where a brief resume of the nature of sound becomes necessary.

SOUND is defined as a disturbance in a material medium caused by the vibration of any body at a certain definite frequency. A sound wave travels through a material medium, such as air or steel, in the form of a compressional wave. This compressional wave travels out from a region of disturbance in exactly the same manner as ripples do when a pebble is dropped into a pool of water. Vibrating objects, such as your vocal cords, cause regions of compressed air, followed by rarefied air, to move outward and away from them in the form of concentric spheres. These vibrations or disturbances reach the ear and cause the eardrum to move inward and outward, according to the pressure exerted by compressions and rarefactions. The human ear is capable of hearing such disturbances only if they occur within the range from 16 to 16,000 cycles per second. The FREQUENCY RESPONSE of the ear is, therefore, said to be from 16 to 16,000 Hz. This range of frequencies is designated by the term AUDIO FREQUENCIES. Although a frequency vibration of 30,000 Hz will cause the diaphragm in the ear to vibrate, the nerves in the ear are incapable of detecting the vibration.

Most of the sound frequencies caused by SPEECH lie between 200 and 3,000 Hertz. Therefore, sound equipment that is used only for voice communication need not be capable of handling audio frequencies beyond this range.

7-6 THE MICROPHONE

An amplifier can only amplify an electrical signal. Therefore, a sound such as music or voice must first be converted into an equivalent electrical signal in order that it may be amplified.

A microphone is a device which translates or converts sound impulses into changing electrical potentials called the signal. The signal, which is now of an electrical nature, can be impressed between the grid and cathode of the first amplifier tube for purposes of amplification.

7-7 THE REPRODUCER

The process of amplification consists of three individual steps:

1. Conversion of sound energy to electrical energy by the microphone.
2. Amplification of the converted electrical energy.
3. Conversion of the amplified electrical energy back into sound energy through the reproducer.

There are two basic types of reproducers in use today that are of interest to the radio amateur. One is the headphone and the other is the loudspeaker. One type of headphone consists of an electromagnet and a metallic diaphragm. The audio energy enters the coil of the electromagnet and causes the diaphragm to vibrate accordingly. The vibrating diaphragm produces sound.

A common type of loudspeaker consists of a permanent magnet and a coil (voice coil) attached to a cone. The audio currents are fed to the voice coil which reacts with the permanent magnet, causing the cone to vibrate. As the cone vibrates, it produces sound.

PRACTICE QUESTIONS — LESSON 7

1 The frequencies that the human ear can hear are:
a. from 30 - 50 kHz
b. 20, 000 - 30, 000 Hz
c. 16 - 16, 000 Hz
d. 16, 000 - 20, 000 Hz

2 A microphone converts sound into:
a. magnetic energy
b. electrical energy
c. direct current
d. vibrations in the air

3 One type of reproducer is:
a. a speaker b. a microphone c. an amplifier d. a magnet

4 An amplifier is defined as a device that
a. converts AC to DC
b. converts DC to AC
c. transforms a small signal into a larger signal
d. uses a crystal pickup

5 A bias supply is called a:
a. speaker b. "B" supply c. "C" supply d. plate circuit

6 The plate supply is called a/an:
a. "B" supply b. "C" supply c. "A" supply d. power source

7 The filament supply is called a/an:
a. "A" supply b. "B" supply c. "C" supply d. "F" supply

8 The input to a triode amplifier is fed to the:
a. plate b. filament c. cathode d. grid

9 The vibrating cone in a speaker:
a. picks up energy from the air
b. transmits energy to the air
c. amplifies sound
d. reproduces RF

10 An RF amplifier will amplify signals whose frequencies are:
a. between 20 and 16,000 Hz.
b. between 30 kHz and 50 kHz.
c. above 10 kHz.
d. above 30 kHz.

SECTION III – LESSON 8
OSCILLATORS

8-1 INTRODUCTION TO TRANSMISSION AND RECEPTION

The first seven lessons of this course were devoted to the study of vacuum tubes, fundamental radio theory and basic circuits. These lessons contain the background material for our discussion of transmitters and receivers. However, before we go into a detailed study of actual transmitter circuits, we will take a bird's eye view of a complete communications system. Instead of drawing out the individual circuits for you, we will draw a series of boxes, each box representing a stage. (A stage is a tube with its associated parts). The function of each stage will be printed inside the box. Such a diagram is known as a block diagram.

Fig. 8-1A illustrates a block diagram of a radio-telephone transmitter. Let us see briefly what the function is of each stage outlined in the block diagram. The heart of the transmitter is the oscillator. Its sole purpose is to

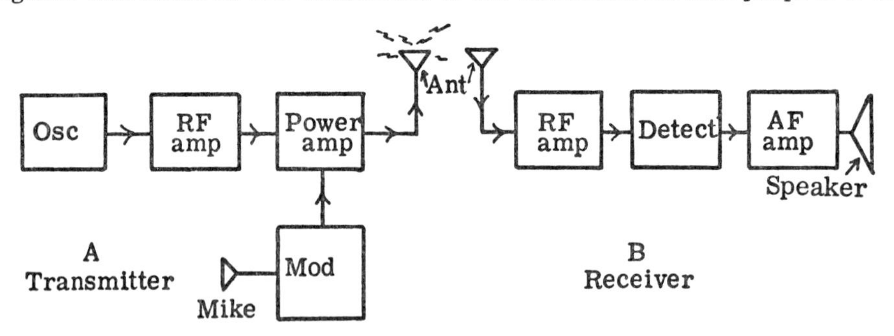

Figure 8-1. Radio transmitter and receiver.

generate a high frequency alternating current. This high frequency AC is called radio frequency or RF. The output of the oscillator is fed to the radio frequency amplifier which simply amplifies the RF output from the oscillator. The output of the RF amplifier is then fed to the RF power amplifier. The RF power amplifier amplifies the RF in terms of power. The power amplifier then supplies the antenna with the RF power that is to be radiated into space. Up to this point, we have only discussed the generation and transmission of a radio frequency wave which, by itself, contains no intelligence. The intelligence that we desire to transmit is the audio in the form of voice or music. Let us see how the audio is radiated into space.

The function of the microphone is to convert sound energy into electrical energy. The output of the microphone is applied to the modulator, which is simply an audio amplifier. The modulator serves two functions: (1) it amplifies the weak audio output of the microphone and (2) it superimposes the audio on to the radio frequency energy that is present at the power amplifier stage by a process called modulation. The audio is combined with the RF wave because an audio wave by itself is not capable of traveling through space. High frequency such as RF, however, is capable of traveling through space. Therefore, the RF acts as the "carrier" for the audio; the RF carries the audio from the transmitter to the receiver. The combined audio RF output of the power amplifier is fed to the antenna where it is radiated out into space in the form of electromagnetic waves.

At the receiving end of the communications system, the electromagnetic waves induce small voltages into the receiving antenna. These signal voltages are quite weak because the electromagnetic waves have travelled some distance before striking the receiving antenna. Therefore, the signal voltages must be amplified; this is the function of the first stage in the receiver called the RF amplifier. The output of this stage is applied to the detector. Just as the oscillator is the heart of the transmitter, the detector is the heart of the receiver. The detector stage separates the audio from the RF carrier. The carrier has served its purpose in bringing the audio to the receiver. Now, all we are interested in is the audio. The audio output of the detector is then fed to an audio amplifier stage to be amplified. The amplified audio is applied to a speaker which converts the audio electrical variations back into the original sound that energized the microphone of the transmitter.

Thus, we have briefly described the overall picture of a communications system. The remaining lessons will go into the details of each stage of a communications system. We will first consider the oscillator of the transmitter.

8-2 INTRODUCTION TO OSCILLATORS

Simply speaking, a vacuum tube oscillator is an electronic alternating current generator. It is a device used to generate an alternating current of any desired frequency. All transmitters and practically all receivers make use of a vacuum tube oscillator. Vacuum tube oscillators are also employed in various types of instruments used for testing and adjusting radio equipment. Because oscillators find so many applications, various types of oscillator circuits have been developed. However, the operation of the different types of oscillators is fundamentally the same.

8-3 THE OSCILLATING TUNED CIRCUIT

The heart of an oscillator is a TUNED CIRCUIT, which consists of a coil and capacitor in parallel. In order to understand how a complete oscillator works, it is first necessary to see how a simple tuned circuit can produce alternating current oscillations. An elementary oscillatory circuit is shown in Figure 8-2. When the switch is thrown to the left, the capacitor "C" is placed across the battery. The coil "L" is out of the circuit. "C" will

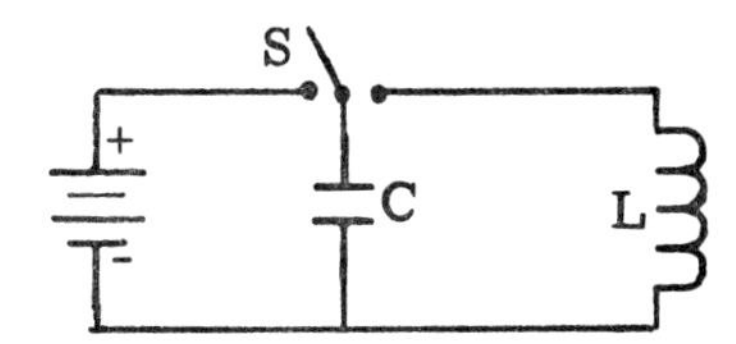

Figure 8-2. An elementary oscillatory circuit.

immediately charge up to the voltage of the battery. The upper plate of "C" will become positive and the lower plate will become negative. A certain amount of electrical energy is, therefore, stored up on the plates of the capacitor by the charging process. If the switch is then thrown to the right, the capacitor will discharge through the coil "L". The electrons will flow from the lower plate of "C", through the coil and back to the upper plate of "C". The flow of electrons will build up a magnetic field around "L". The energy which was stored in the capacitor has been transferred over to the magnetic field surrounding the coil. When "C" is discharged completely, the flow of electrons through "L" tends to cease, causing the mangetic field to start collapsing. The collapsing magnetic field induces a voltage of such a polarity across "L" that it maintains the flow of electrons to the upper plate of the capacitor. This occurs because the magnetic field acts to prevent a change in the flow of current. The flow of electrons to the upper plate continues until it is negative with respect to the bottom plate. When the magnetic field has completely collapsed, the energy which was in the magnetic field has been transferred over to the capacitor in the form of a stored charge. The capacitor is now charged in the opposite polarity to its original charge. The capacitor again discharges through "L", and the entire action as outlined above repeats itself. Thus we can see that the current OSCILLATES back and forth between the coil and the capacitor, alternately charging "C", first in one direction and then in the other. This alternating current will produce an alternating voltage across the tuned circuit. The frequency of this AC voltage is determined by the values of "L" and "C".

If there were no resistance in either the coil or the capacitor, there would be no energy loss in the form of heat. The oscillations would, therefore, continue forever at a constant amplitude. However, such a situation is impossible in actual practice. Some resistance is always present in radio components, especially in a coil. This resistance causes some of the energy which oscillates back and forth in the tuned circuit to be transformed into heat. The heat, of course, is a loss of energy. Therefore, with each succeeding cycle, the amplitude of the oscillating voltage decreases until all of the energy has been dissipated.

8-4 CONDITION FOR OSCILLATION

In radio, it is necessary that the tuned circuit oscillations continue at a constant amplitude. If we want the oscillations to continue, we must make up for the resistance losses which occur in the L-C circuit. We must somehow inject electrical energy back into the L-C circuit to sustain the oscillations. Where is this energy to come from and how do we inject it properly into the L-C circuit? To clarify this question in our mind, we can compare the oscillations of energy in the tuned circuit to a child on a swing. In order that the child keep swinging at a constant height, it is necessary that someone give the swing a little push each time the child reaches the top of his swing. In other words, energy must be added to the swing at the right time and of the right amount to overcome the friction in the hinges, otherwise the swing will gradually come to rest, just like the damped wave oscillations.

In radio, the answer to the question of how to maintain oscillation lies in the use of the amplifying ability of the electron tube or transistor. If the oscillating circuit of Figure 8-2 were connected to the grid circuit of a vacuum tube, an amplified version of the oscillating voltage would appear in the plate circuit. If we could somehow continuously feed back some energy from the plate circuit to the grid circuit to compensate for the resistance losses in the L-C grid circuit, oscillations could continue. A simple method of doing this is shown in Figure 8-3. L_1 and C_1 represent the tuned circuit, sometimes called the TANK CIRCUIT. V_1 is the triode amplifier tube. L_p is a coil of wire wound on the same form and next to L_1. Since L_p is in the plate circuit, it is easy to see that some of the amplified energy from the plate circuit is fed back to the grid circuit through the magnetic coupling between the two coils. This energy will overcome the losses in the tank circuit, and oscillations will be maintained.

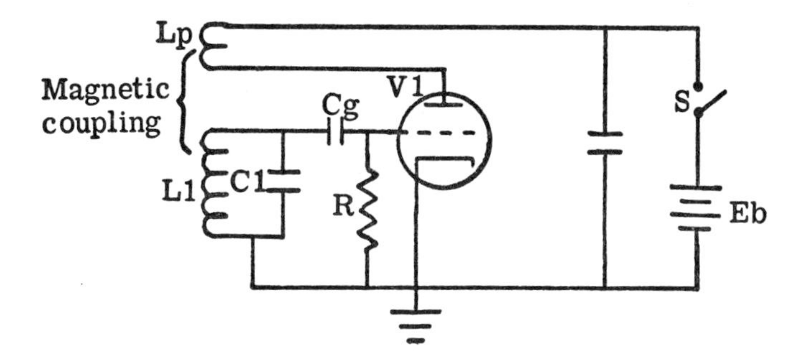

Figure 8-3. Tickler-coil oscillator or Armstrong oscillator.

The entire circuit of Figure 8-3 is called a vacuum tube oscillator. This particular oscillator has found wide practical use. It is known by the names of: TUNED GRID OSCILLATOR, TICKLER COIL OSCILLATOR, or ARMSTRONG OSCILLATOR.

Figure 8-4 illustrates the Armstrong oscillator using a transistor instead of a tube.

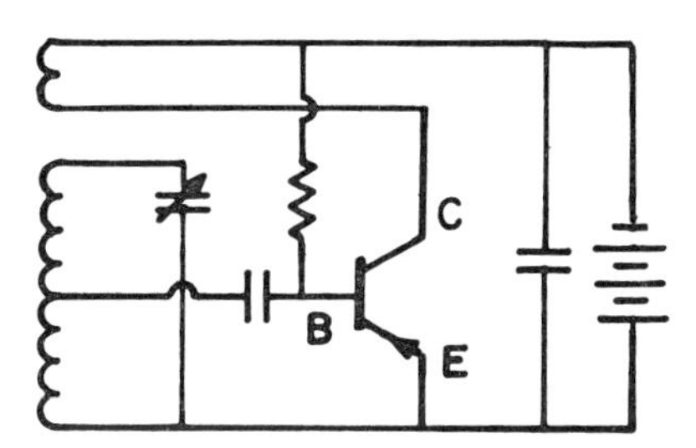

Figure 8-4. Armstrong oscillator using a transistor.

8-5 FREQUENCY STABILITY

Most electronic components are effected by changes in temperature. Resistors, capacitors, and inductors as well as the characteristics of vacuum tubes and transistors will change in value when subjected to changes in temperature. In most electronic circuits, these changes in value are not critical and will not unduly affect the performance of the circuit. However, when electronic components are used in frequency determining circuits, changes in electrical characteristics and values also change the operating frequency of the circuit. When you are communicating with another amateur on a given frequency, he will only hear you if your transmission remains on the frequency to which he is tuned. The characteristic of a transmitter to remain on frequency is referred to as FREQUENCY STABILITY and the characteristic of frequency instability is referred to as frequency drift. The most critical circuit affecting frequency stability in a transmitter is the oscillator.

8-6 CRYSTAL-CONTROLLED OSCILLATORS

The most stable of all oscillators is the CRYSTAL-CONTROLLED OSCILLATOR. It has a much greater frequency stability than the Armstrong oscillator. The difference between the Armstrong oscillator and the crystal oscillator is that the oscillator tuned circuit consisting of L and C is replaced by a crystal substance. This crystal is usually made out of quartz, a mineral found in the earth. The quartz crystal has the following peculiar property. If a mechanical vibration is applied to the quartz crystal, an electrical voltage will be developed across its surfaces. On the other hand, if we apply an alternating voltage to the surfaces of the quartz crystal, it will vibrate mechanically at the frequency of the AC voltage. This property of the quartz is known

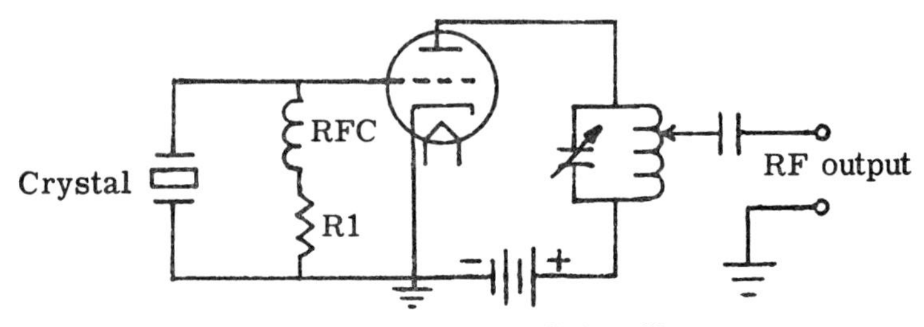

Figure 8-5. Crystal-controlled oscillator.

as the PIEZO-ELECTRIC EFFECT. The crystal, from an electrical viewpoint, acts in the same manner as a tuned circuit. If energy is injected into a crystal, an electrical oscillation is generated across the crystal surface, which continues until all of the energy has been used up. Since the vibrating crystal is similar to a tuned circuit, it can be placed in the grid circuit of a tuned plate-tuned grid oscillator, in place of the actual tuned grid circuit. A schematic of a triode crystal oscillator is shown in Figure 8-5. Energy from the plate tuned circuit is fed back to the grid circuit through the grid-plate capacitance of the tube. The energy that is fed back to the grid circuit keeps the crystal oscillating. The oscillations occur at the resonant frequency of the crystal, and the plate circuit is tuned approximately to this frequency. The resonant frequency of a crystal is determined primarily by its physical dimensions.

PRACTICE QUESTIONS — LESSON 8

1 An oscillator is a/an:
a. generator   b. amplifier   c. rectifier   d. reproducer

2 The basic components of a tuned circuit are:
a. a coil and resistor in series
b. a resistor and capacitor in series
c. a coil and capacitor in parallel
d. a resistor and capacitor in parallel

3 The most stable of all oscillators is the:
a. tuned circuit   c. crystal oscillator
b. variable frequency oscillator   d. Armstrong oscillator

4 In an oscillator, a crystal is used in place of the

a. L-C circuit
b. choke
c. load resistor
d. coupling network

5 The resonant frequency of a crystal is controlled by its:
a. weight
b. physical dimensions
c. feedback
d. input voltage

6 A transmitter does not contain a/an:
a. oscillator
b. detector
c. RF amplifier
d. Modulator

7 A detector is found in a:
a. receiver.
b. modulator.
c. R.F. amplifier
d. oscillator.

8 An important condition for oscillation is:
a. high plate voltage
b. low plate voltage
c. feedback
d. high gain

SECTION III – LESSON 9
CONTINUOUS-WAVE TRANSMITTERS

9-1 INTRODUCTION

The function of a radio transmitter is to transmit intelligence by means of a radio frequency wave. The RF wave is radiated into space by an antenna system. An antenna is a device which converts AC energy into electromagnetic radiation known as radio waves. The RF wave traveling through space is then picked up by a receiver which converts the RF signal into an audio output.

Radio transmitters may be divided into two types. One is the CONTINUOUS-WAVE type of transmitter, which we shall now study; the other is the modulated type of transmitter, which we shall study later on.

9-2 CONTINUOUS WAVES

Continuous waves, abbreviated CW, are radio waves of constant amplitude. In the CW transmitter, continuous waves are radiated into space by simply coupling the output of a vacuum tube power oscillator to a suitable antenna system. The International Morse Code is used to convey intelligence by CW communication.

The Morse Code consists of a series of dots and dashes which represent the letters of the alphabet. In order to transmit code, the CW transmission must be interrupted in a dot and dash sequence. This type of emission is actually an RF wave broken up into sections. An oscillator is made to stop and start oscillating by means of a telegraph key. By allowing the oscillator to operate for longer or shorter amounts of time, we can produce dots and dashes. Figure 9-1 shows the output of an oscillator for the letter "D" (dash-dot-dot).

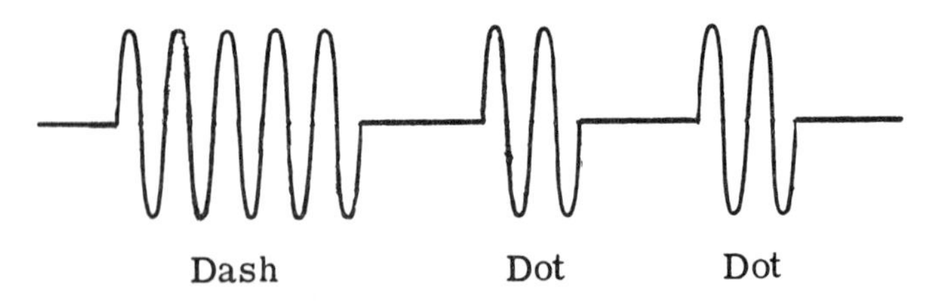

Figure 9-1. Keyed output of an oscillator for the letter "D" (dash-dot-dot).

The symbol that is used to designate the transmission of code by interrupting a CW transmitter is "A1". The symbols for the various types of transmission can be found on Page 78

9-3 ONE-TUBE TRANSMITTER

In early type radio transmitters, the oscillator was directly coupled to the antenna system. In order to increase the power output of this type of transmitter, it was necessary to use a larger tube or to increase the operating voltages. There is a limit, however, to the amount of power that one can get from a one-tube transmitter. The power output of an oscillator depends upon RF currents in the oscillator circuit. Since these currents are relatively weak, very little power can be delivered to the antenna. The radiated wave, therefore, will also be weak. Another defect of the simple oscillator type of transmitter is its poor frequency stability. Figure 9-2 shows a one-tube transmitter. Capacitor C_A represents the antenna capacitance to ground which will vary as the antenna swings in the wind. This varying antenna capacitance will be coupled back to the tank circuit and will cause the oscillator frequency to vary. The disadvantage of poor frequency stability can be overcome to a great extent by the use of an intermediate amplifier stage which serves to isolate the antenna from the oscillator. Changes in antenna capacity will therefore, not be reflected back into the oscillator tank circuit. At the same time, the amplifier amplifies the output of the oscillator and feeds a more powerful signal into the antenna.

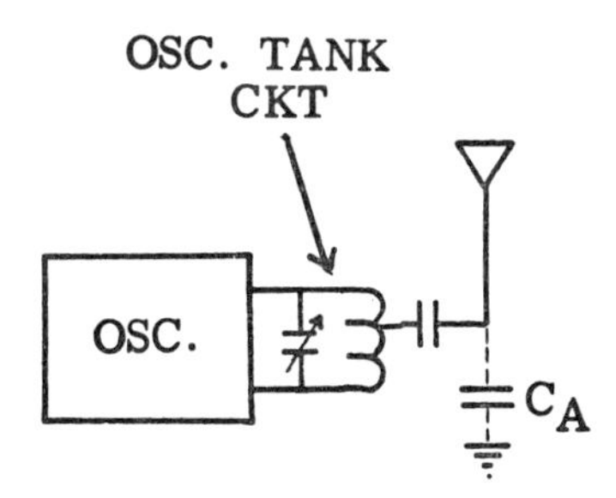

Figure 9-2. One-tube transmitter.

9-4 MASTER-OSCILLATOR POWER-AMPLIFIER

A transmitter consisting of an oscillator and an amplifier (or a series of amplifiers) is called a MASTER-OSCILLATOR POWER-AMPLIFIER, MOPA for short. Such a transmitter is shown in Figure 9-3. The output of the oscillator is amplified by V_2. Capacitor C_1 prevents the high DC voltage on the plate of V_1 from being applied to the grid of V_2. At the same time, it allows the RF energy to get through to the grid of V_2. The RF choke L_1 prevents the RF energy from flowing to ground through R_1. This is because an RF choke opposes the flow of RF currents.

The master-oscillator power-amplifier type of transmitter has a decided advantage over the simple oscillator transmitter in that the frequency stability is greatly improved. High frequency stability is obtained in this system because the oscillator is not coupled directly to the antenna; the oscillator is, therefore, unaffected by any change in the antenna-to-ground capacitance. Changes in antenna-to-ground capacitance will merely react upon the RF power amplifier circuit, resulting in a decrease in the radiated power output. The amplifier in Figure 9-3 may feed the antenna directly, or it may be the first of a series of RF amplifiers, the last of which feeds into an antenna system.

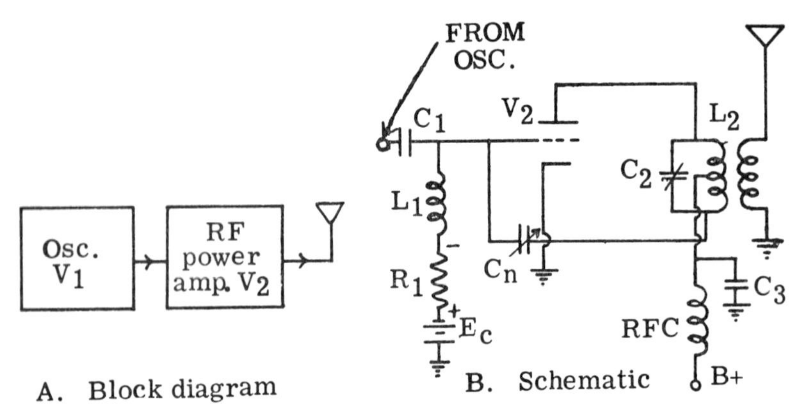

Figure 9-3. Master-oscillator, power-amplifier transmitter.

9-5 KEYING THE TRANSMITTER

At the beginning of this lesson, we learned that the telegraph key is used to start and stop the operation of a CW transmitter. This causes radio waves to be sent out by the antenna in the form of dots and dashes. The telegraph key is merely a switch which opens and closes a circuit or circuits in a transmitter.

While keying may at first thought seem simple, there are many considerations which make it a very important subject for study. These considerations concern not only the simple act of forming dots and dashes by keying, but also the undesired effects that may result from interrupting the operation of the transmitter.

A good keying system should fulfill the following requirements:

1. There should be no radiation of energy from the antenna when the key is open (key up). Some energy may get through to the antenna during keying spaces (when the key is open). The energy that is radiated during key-up is called BACKWAVE. A CW signal containing backwave is very difficult to read because a weak signal is heard in the receiver during the space interval between dots and dashes. This signal may be almost as loud as the code reception. A pronounced backwave often results when the keying is done in the amplifier stage feeding the antenna. Backwave may also be caused by incomplete neutralization of the final stage. This allows energy to get to the antenna through the grid-plate capacitance of the tube. A third cause of backwave may be the possible magnetic pickup between the antenna coupling coils and one of the low power stages. Backwave can generally be eliminated by shielding, by proper neutralization, and by rearranging the tank circuits to eliminate unwanted coupling.

2. The keying system should allow the radiation of full power output when the key is closed (key down).

3. The code output should be free of clicks.

When power is applied or removed from a circuit very suddenly, as is the case when a transmitter is keyed, the large amounts of energy that are thus released will surge back and forth and will result in damped oscillations. The damped oscillation will cause interference in nearby receivers. Interference will be present in the form of clicks or thumps, even though the receivers are tuned to different frequencies from that of the transmitter. KEY CLICK FILTERS are used in the keying system of radio transmitters to attenuate or reduce the clicks. A typical key click filter is shown in Figure 9-4.

The inductance, L, causes a slight lag in the current which builds up gradually instead of instantly when the key is closed. C and R are connected in series across the key to absorb the spark which tends to occur when the key is opened. The capacitor charges up and prevents a spark from jumping the gap formed by the open key. When the key is closed, the capacitor discharges through R, thereby dissipating the energy of the charged capacitor.

4. The code output should be free of chirps. Chirps occur because the transmitter's output frequency changes when the key contacts are closed and when they are opened. It is difficult to prevent chirps completely because the sudden voltage changes that occur during keying will affect the frequency.

Some of the means by which chirping can be minimized are: use a regulated power supply on the oscillator, key an amplifier rather than the oscillator, use a buffer stage between the keyed amplifier stage and the oscillator stage, use a separate power supply for the oscillator.

5. The code output should be free of hum. If the filter of a transmitter power supply is not working properly, or if there is inadequate filtering, an AC hum will be superimposed on the carrier. The output of the receiver will reproduce this hum. The hum can be eliminated if the defective filter component is replaced.

9-6 METHODS OF KEYING

Keying takes place in either the oscillator or amplifier stages of the transmitter. A number of different keying systems are in use today. Figure 9-5 illustrates PLATE KEYING. Plate keying may be used in the oscillator or amplifier stages. When the key is open, no plate current can flow and the tube does not operate. When the key is closed, the tube operates and the transmitter sends out RF. The key may be used to control the plate current of one tube or several tubes. Plate keying is usually accomplished in the power amplifier circuit in transmitters which use a crystal controlled oscillator.

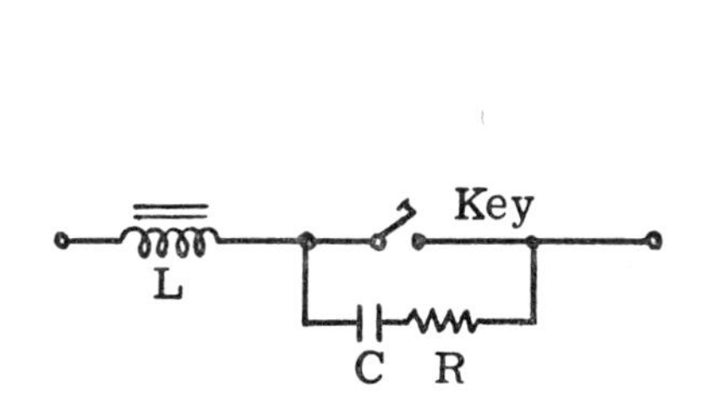

Figure 9-4. Key click filter.

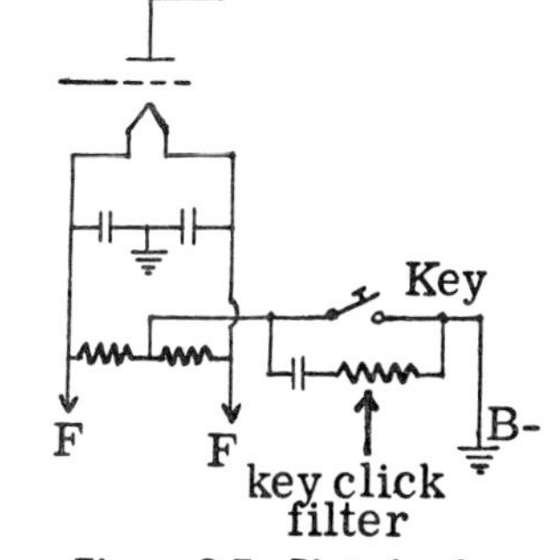

Figure 9-5. Plate keying.

In larger transmitters, the ordinary hand key cannot accommodate the large plate current flow without excessive arcing. The high plate voltage may also make it too dangerous to operate a hand key in the plate circuit. Therefore, some indirect method of stopping and starting the plate current is called for. Figure 9-6 shows a relay system for indirectly controlling the plate current.

A widely used method of keying is called BLOCKED-GRID KEYING. In this method, a high negative bias is applied to the grid of the stage to be keyed. This bias is sufficient to cut off plate current completely, even with

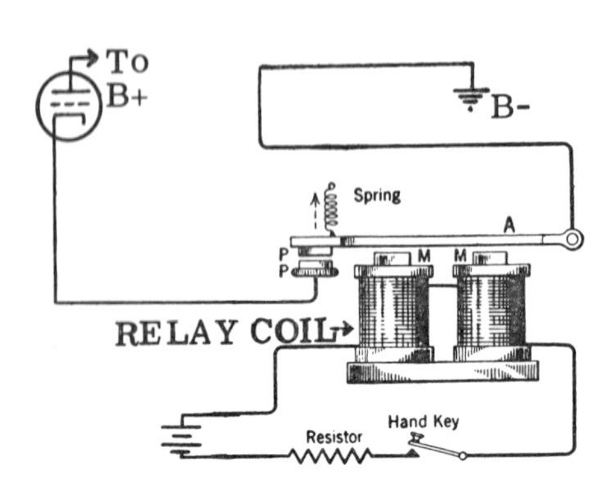

Figure 9-6. Indirect keying using relay.

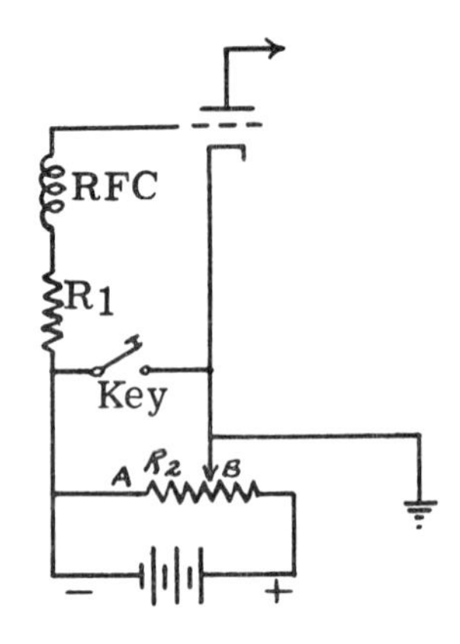

Figure 9-7. Blocked-grid keying

excitation applied to the grid. Figure 9-7 illustrates the method of blocked-grid keying. R_1 is the normal grid-leak resistor. R_2 is a voltage dividing resistor across which the entire voltage is placed. When the key is up, the large negative voltage across the AB portion of R_2 is applied to the grid. This voltage is large enough to cut the tube off completely, regardless of the size of the input signal. Therefore, the stage is inoperative and there is no output from the transmitter. When the key is depressed, the blocking bias across AB is shorted out, and the only bias in the stage is the grid-leak bias. This is the normal bias for the stage, and the transmitter now radiates its normal output.

9-7 INPUT POWER LIMITATIONS

Paragraph 97.67(d) of the Federal Communications Commission's Rules and Regulations states as follows: "in the frequency bands 3700 - 3750 kHz, 7100 - 7150 kHz (7050 - 7075 kHz when the terrestrial location of the station is not within Region 2), 21,100 - 21,200 kHz, and 28,100 - 28,200 kHz, the maximum power input to the transmitter final amplifying stage supplying radio frequency energy to the antenna shall not exceed 250 watts, exclusive of power for heating the cathode of a vacuum tube(s)." Since these frequencies are those that a Novice may use, the maximum Novice input power is 250 watts. Also, since this rule does not limit the input power to the PLATE input power, we must add up all the power inputs to all the elements of the tube or transistor, with the exception of the power used for the filament or cathode. In other words, if the final stage of a Novice transmitter used a tetrode vacuum tube, the input power would be equal to the sum of the plate input power, the screen grid input power, and the driving power on the control grid. An example will clarify this.

EXAMPLE: What is the input power to the final RF stage of a Novice transmitter having the following voltages and currents?

final plate voltage of 600 V.
plate current of 80 ma.
screen voltage of 150 V.
screen current of 8 ma.
filament voltage of 6.3 V.
filament current of 500 ma.
driving power of .6 watts.

SOLUTION: We first determine the plate and screen powers using the basic power formula: P = EI.

Plate input = P = E x I = 600 x .08 = 48 watts.
Screen input = P = E x I = 150 x .008 = 1.2 watts.

To arrive at the power input to the stage, we add up the plate power, the screen power and the driving power. We disregard the filament power.

Input power = 48 + 1.2 + .6 = 49.8 watts.

9-8 MINIMIZING HARMONIC OUTPUT

The output signal of a transmitter should be a pure sine wave at the frequency at which we are transmitting. In other words, if we are transmitting a signal at 7,120 kHz, we should be radiating only 7,120 kHz and nothing else. However, this is not always the case. If we were to analyze the output of the transmitter, in addition to transmitting a 7,120 kHz signal, we also find that we are transmitting some energy at 14,240 kHz, and perhaps a slight amount of energy at 21,360 kHz. The 14,240 kHz signal and the 21,360 kHz signal are known as HARMONICS. Harmonic frequencies are whole numbered multiples of the fundamental frequency. The second harmonic is twice the fundamental frequency, the third harmonic is three times the fundamental frequency, the fourth harmonic is four times the fundamental frequency, etc. In the above example, 7,120 kHz is the fundamental frequency. 14,240 kHz is twice the fundamental frequency and therefore, is the second harmonic. 21,360 kHz is three times the fundamental frequency and is, therefore, the third harmonic.

It is very important that the output waveform of a transmitter contain a minimum of harmonic components. If the harmonic components are radiated by the antenna, unlawful interference may result. There are many methods of minimizing the harmonic output of a transmitter. Some are given below, others will be given in some of the paragraphs and lessons later in the book.

The final RF amplifier should have the following characteristics if the generation of harmonics is to be kept to a minimum:

(1) The amplifier should operate with proper grid bias and grid signal drive.

(2) Avoid capacitive coupling.

(3) Use tuned circuits where possible.

(4) Use proper plate and screen voltages.

In addition to producing harmonic signals, a transmitter's output may contain signals that are due to undesired oscillations. The undesired harmonics, and undesired oscillations, as well as key clicks and chirps are sometimes referred to as SPURIOUS SIGNALS or SPURIOUS EMISSIONS.

9-9 THE PI-SECTION OUTPUT CIRCUIT

The Pi-section output circuit is in widespread use in both commercial and amateur transmitters. It provides a certain amount of attenuation of undesired harmonics and it permits the transmitter to be used with a variety of antennas.

Fig. 9-8 illustrates an RF stage using a pi-network output circuit. The pi-network circuit consists of C3, L and C4. This circuit is not too different from the conventional parallel resonant output circuit. C3 is the plate tuning capacitor and C4 is the loading capacitor.

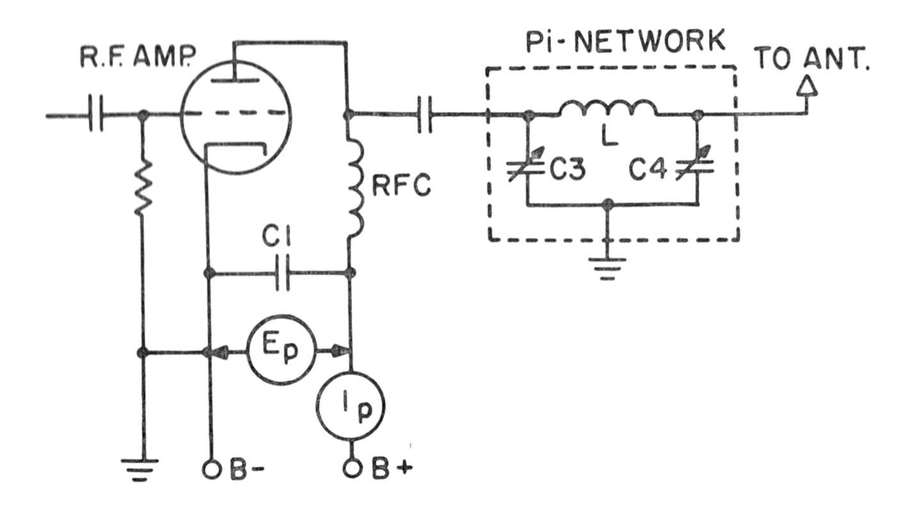

Figure 9-8. An RF amplifier stage using a pi-network output tank circuit. Neutralization is not shown in this diagram.

The proper tune-up procedure for the pi-network output circuit is as follows:

(1) Before power is turned on, adjust C3 and C4 for maximum capacitance.

(2) After the power has been turned on and the circuits ahead of the RF amplifier stage have been tuned up, adjust C3 for minimum reading (dip) on the plate current meter.

(3) Reduce the capacitance of C4 slightly and observe that the plate current rises.

(4) Readjust C3 for a dip in the plate current.

(5) Reduce the capacity of C4 again and note a further rise in plate current.

(6) Readjust C3 for a dip in plate current.

(7) Repeat steps 5 and 6 until the plate current reading is at its recommended value at the "dip" point. C3 should be the last control to be adjusted.

9-10 RADIO FREQUENCY INTERFERENCE (RFI)

Radio Frequency Interference (RFI) is the interference caused to other equipment by the RF emissions from a transmitter. Amateur stations can cause RF interference to a number of other services. Some of them are: television receivers, standard broadcast receivers, hi-fi equipment and the telephone system. This consumer equipment is usually located close to the amateur transmitter and is affected by the transmitter's strong RF field.

Television interference (called TVI) is the most important type of interference confronting the amateur. It is discussed below in detail. Interference to broadcast receivers and hi-fi equipment is usually due to rectification of the strong RF signal in an early audio stage. This can be cured by shielding and/or bypassing of power leads, speaker leads and other interconnecting leads. By-passing the grid of the first audio tube to the cathode with a .001 mfd. capacitor is important in eliminating this type of interference.

Pickup of the transmitter signal in the telephone lines can usually be cured by RF by-passing of the telephone microphone and their lines.

RFI in P-A systems is handled in a manner similar to that of hi-fi equipment.

9-11 TELEVISION INTERFERENCE

Amateur transmitters frequently cause interference to television receivers. One of the main reasons for this is the fact that the frequencies of the television channels are harmonically related to the amateur band frequencies. An example will make this clear. Let us assume that an amateur is operating at a frequency of 29 MHz. Regardless of how well the transmitter is designed, a certain amount of energy at the second harmonic of 58 MHz. (2 x 29 MHz) will also be transmitted. Since television Channel 2 is from 54 to 60 MHz, the 58 MHz signal will cause interference to Channel 2.

One of the important methods of preventing harmonics from being radiated from the transmitter is to install a low-pass filter between the transmitter and its antenna. A low-pass filter will pass signals below a certain cut-off frequency and block signals above that frequency. In this way, the harmonics will not get through to the antenna to be radiated. In the above example, we would install a low-pass filter with a cut-off frequency of 40 MHz. The 29 MHz signal would get through, but the 58 MHz harmonic would be suppressed.

Figure 9-9 illustrates some simple forms of single section low-pass

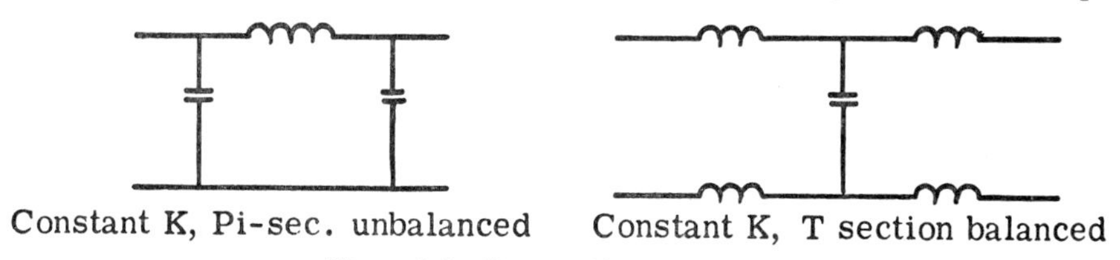

Figure 9-9. Types of low pass filters.

filters, together with the designations of the filters. The designations come about from the fact that the appearance of the various filter configurations resemble letters of the American or Greek alphabet.

Other methods of suppressing spurious and harmonic emissions from the transmitter are as follows:

(1) Use a Transmatch or Antenna Tuner between the transmitter and the transmission line. Fig. 9-10 is a block diagram showing where the filter and tuner are inserted in a transmitter installation.

(2) Use a transmitter circuit design and layout that will not cause harmonic and spurious signals in the TV bands.

(3) Where possible, use link coupling and tuned circuits.

(4) Shield the transmitter adequately. Ground all cabinets to a good common ground such as the cold water pipe or a ground rod.

(5) Keep plate and screen voltages as low as possible.

(6) Reduce excessive grid drive to the final stage.

(7) Check all mechanical joints and connections in the antenna system and ground system. Make sure that they are tight and free of corrosion.

The above paragraphs deal with television interference caused by spurious and harmonic output from the transmitter. Television interference can also be caused by receiver deficiencies. A receiver that is not highly selective will respond to interfering signals outside of its frequency band.

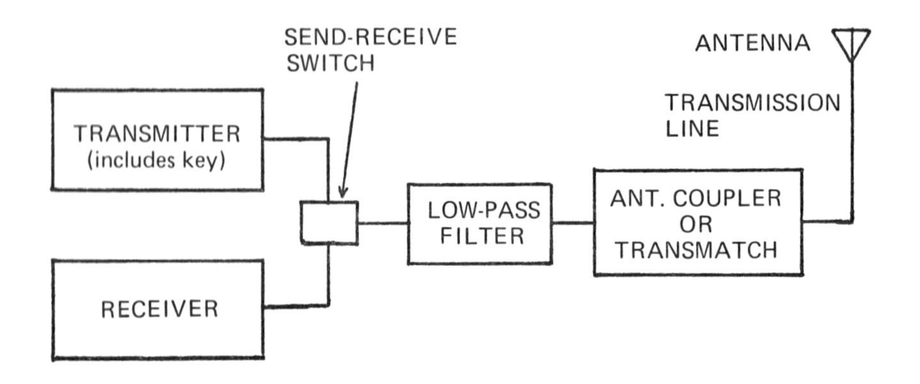

Figure 9-10. Block diagram of a radio station.

This is usually true of TV receivers located in the vicinity of a transmitter. For instance, let us assume that an amateur, operating at 29 MHz, has adequate harmonic suppression and transmits a clean signal. A TV receiver, located close by, will be swamped with the 29 MHz signal. It will overload the receiver's front end and by means of rectification in the receiver, harmonics will be produced. The second harmonic of 29 MHz is 58 MHz. This falls inside of Channel 2 and will cause interference as though the amateur transmitter were transmitting a 58 MHz signal.

The way to prevent this type of interference is to install a high-pass filter at the receiver's antenna terminals in series with the antenna. A high-pass filter will pass signals above a certain cut-off frequency and reject signals below this frequency. We choose a high-pass filter with a cut-off frequency below the television frequencies. For instance, 40 MHz could be used as the cut-off frequency. The television signals will pass through without attenuation, but the signals below the cut-off frequency will not be allowed to pass. In the above example, the 29 MHz Amateur signal would be attenuated while the TV signals, which are above 50 MHz would pass without attenuation. Some typical single section high-pass filters are illustrated in Fig. 9-11. Filters with greater attenuation can be made by combining two or more single section filters in series.

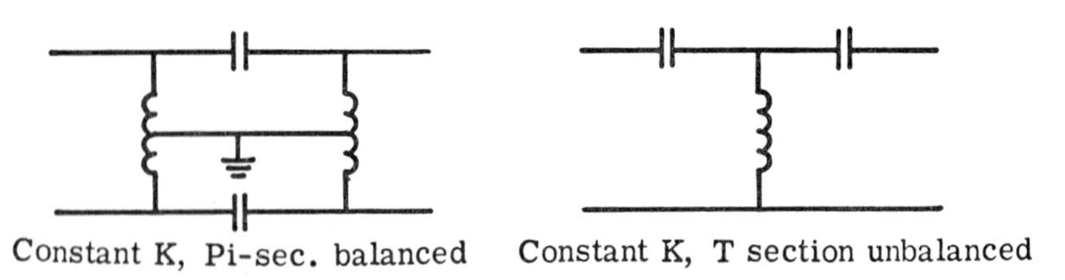

Figure 9-11. Types of high-pass filters.

Other methods of reducing TVI at the receiver end are:

(1) Use wave traps tuned to the amateur transmitting frequency at the antenna and/or the power lines.

(2) Use an AC line filter.

(3) Use a high gain TV receiving antenna.

(4) Use a shielded type of transmission line.

(5) Check the receiving antenna for broken elements, corroded joints and poor electrical and mechanical connections. Corroded joints and poor electrical connections can behave as rectifiers, causing harmonic generation.

9-12 SAFETY

High voltages are present in transmitters. They are dangerous and can cause severe electrical shock, or even death. The following precautions should be observed when working with high voltage circuits:

1. Shut off the power from circuits that you intend to work on.
2. Even if the transmitter has a door interlock switch, check high voltage points with a meter to make sure that the door interlock switch has actually removed the B+ when the door was opened.
3. Short a B+ point to ground with an insulated screw driver to discharge the energy that may remain in the capacitors.
4. If the plate voltage of the final amplifier has to be measured, the meter should be placed between the B+ point and the cathode (or ground) rather than between the plate and cathode. The plate has a high RF voltage on it and there is the danger of arcing or RF burns if anything touches the plate.

In designing electronic equipment that will have high voltages, use bleeder resistors across the power supply filter capacitors. Furthermore, high voltage wires or circuits should not be exposed or in a position where an operator or other person can come in contact with them.

All electronic equipment having high voltage, should be enclosed in metal cabinets. The cabinets should be connected to each other and to a good ground. By a good ground, we mean the cold water system or a ground rod. A description of a ground rod is given in Lesson 12.

9-13 EMERGENCY TREATMENT FOR ELECTRIC SHOCK

It is extremely important to observe the safety precautions outlined above. Members of your family and others around you should also be taught to observe all the safety precautions. However, accidents do occur. If someone does receive a severe electric shock, the following emergency treatment should be given:

1. If the victim is still in contact with a live circuit or live wire, immediately turn off the switch that controls the current to the live circuit. If this isn't possible, take a long dry wooden pole or other insulated object and push the wire away from the victim. Do these things WITHOUT ENDANGERING YOURSELF.
2. Send for medical aid AS SOON AS POSSIBLE. If another person is around, have him send for medical aid at the same time that you are doing step 1. Step 2 is extremely important. The sooner that competant medical aid is given to the victim, the greater are his chances of survival.
3. Determine whether the victim is breathing. If he is, loosen his clothing so that he can breathe easily. Keep him lying down in a comfortable position. Protect him from exposure to cold.
4. If the victim is not breathing, apply artificial respiration.

PRACTICE QUESTIONS — LESSON 9

1 In a pi-network output circuit, the input capacitor should be:
   a. tuned for minimum antenna current
   b. tuned for maximum plate current
   c. tuned for minimum plate current
   d. a and c are correct

2 The principle purpose of using door interlock switches is to:
   a. eliminate the need of turning off transmitter
   b. act as an on-off switch
   c. protect equipment against mishandling by incompetent personnel
   d. prevent personnel from being accidentally shocked by dangerous voltages when cage to transmitter is open

3 TVI can be eliminated by a:
   a. high-pass filter at the receiver
   b. high-pass filter at the transmitter
   c. high-pass filter at the receiver and a low-pass filter at the transmitter
   d. low-pass filter at the receiver

4 A Pi-network allows a transmitter to be used with a wide variety of:
   a. receivers   b. microphones   c. antennas   d. amplifiers

5 Low pass filters are used to:
   a. Reduce TVI
   b. couple the transmitter to the antenna
   c. increase power
   d. reduce SWR

6 The maximum input power for a novice station is:
   a. 75 watts   b. 100 watts   c. 150 watts   d. 250 watts

7 Which of the following is not a method of keying a transmitter:
   a. plate keying
   b. filament keying
   c. transformer primary keying
   d. blocked grid keying

8 By-passing the grid of the receiver's first audio tube with a small capacitor will prevent:
   a. radio frequency interference
   b. oscillation
   c. regeneration
   d. grid saturation

9 Interference to nearby receivers due to keying large amounts of energy in a transmitter can be eliminated by a:
   a. back wave filter
   b. low pass filter
   c. key click filter
   d. high pass filter

10 The addition of an RF power amplifier to a one-stage CW transmitter does not increase:
   a. frequency stability
   b. output power
   c. number of stages
   d. antenna capacity

11 In the event of a severe electrical shock, which of the following is NOT an initial step?
   a. Send for a doctor
   b. apply artificial respiration
   c. shut off power
   d. break victim's contact with line circuit

12 What is the third harmonic of 3,000 kHz.?
a. 27 kHz b. 9 MHz c. 1 MHz d. 100 kHz

13 What is the power input to the final RF stage of a Novice transmitter having a final plate voltage of 400 V., a plate current of 50 ma., a screen voltage of 300 V., a screen current of 10 ma., a filament voltage of 6.3 V., a filament current of 500 ma., and a grid drive power of .5 watts?
a. 20 watts b. 26.65 watts c. 23 watts d. 23.5 watts

14 A Transmatch will match impedances between the:
a. transmitter and the antenna system.
b. receiver and the antenna.
d. transmitter and the receiver.
d. transmission line and the antenna.

15 A key-click filter contains:
a. a capacitor and resistor in series, both across the key
b. two capacitors in series, both across the key.
c. a capacitor and inductor in series, both across the key.
d. an inductor and resistor in series, both across the key.

16 In blocked grid keying:
a. a large bias is placed on the tube when the key is up.
b. a high voltage is placed on the plate when the key is up.
c. a low voltage is placed on the cathode when the key is up.
d. a low voltage is placed on the plate when the key is up.

17 What is the power to the final stage of a Novice transmitter, having a plate voltage of 600 volts, a plate current of 125 ma., a filament voltage of 12 volts, a filament current of 300 ma., a grid drive power of .8 watts?
a. 79.4 watts. b. 72.2 watts. c. 75 watts. d. 75.8 watts.

SECTION III – LESSON 10
THE MODULATED TRANSMITTER

10-1 INTRODUCTION

In the previous lesson, we learned how a CW transmitter operates. Communication by means of CW code transmission is known as RADIOTELEGRAPHY. The disadvantage of radiotelegraphy transmission is that the radio operator must know code. In order that operators who are not familiar with code be able to send and receive messages directly, the transmission of speech is necessary. The transmission of audio (speech) by means of radio communication is known as RADIOTELEPHONY.

A radiotelephone transmitter consists of a CW transmitter (minus the telegraph key), plus an audio frequency amplifier system. The audio frequency system amplifies the audio signals and superimposes them on the RF signal that is generated by the RF oscillator. The process of superimposing the audio on the RF is known as MODULATION. The RF signal is called a CARRIER since it "carries" the audio through space to the receiving antenna. The frequency of the signal is called the CARRIER FREQUENCY.

10-2 AMPLITUDE MODULATION

There are several methods of modulating a carrier. One of the basic methods is called AMPLITUDE MODULATION.

In amplitude modulation, the modulating frequency is the intelligence (voice or music) which is to be transmitted through space to receivers many miles away. The modulating frequency is audio and, by itself, cannot be transmitted. A radio-frequency wave, however, is capable of being transmitted through space. If we combine or mix an audio-frequency wave with a radio-frequency wave in a special mixing circuit, we obtain an RF output which contains the audio and can be transmitted. Figure 10-1 illustrates a

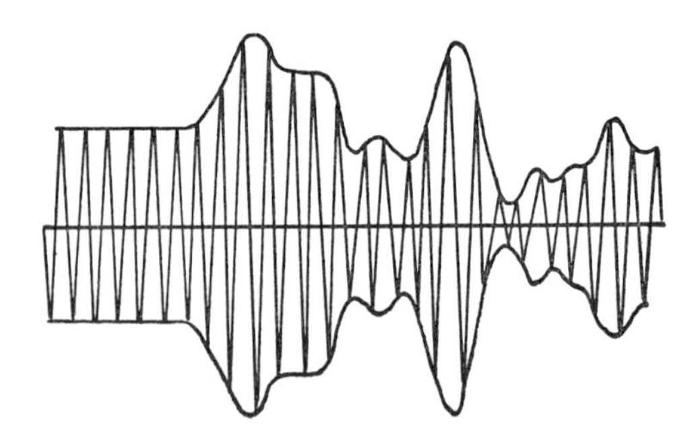

Figure 10-1. Radio wave modulated with voice.

voice modulated radio frequency wave whose amplitude varies according to the amplitude of the audio wave, thus the term "amplitude modulation." The frequency of this variation is the same as the audio modulating frequency. (The abbreviation for amplitude modulation is AM). An AM wave is, therefore, a radio-frequency wave which contains in its amplitude variations, the audio or intelligence which we desire to transmit.

10-3 THE AM TRANSMITTER

A block diagram of a typical amplitude modulated radiotelephone transmitter is shown in Figure 10-2. Above each block is drawn the waveshape of the voltage output of that particular stage. With the aid of these waveshapes and the block diagram layout, we shall discuss the operation of the radiotelephone transmitter.

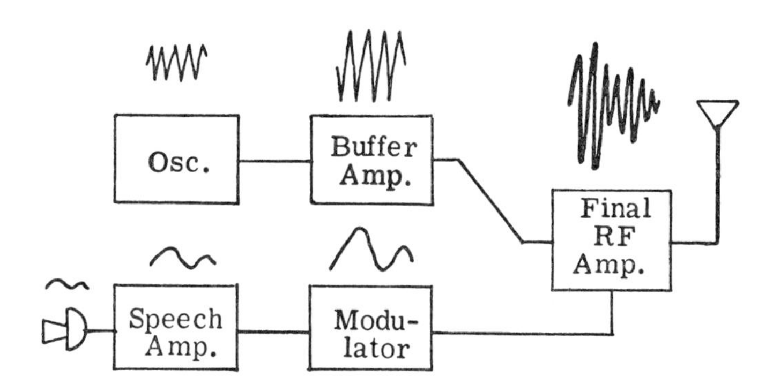

Figure 10-2. Block diagram of amplitude-modulated transmitter.

To begin with, the oscillator stage generates a radio-frequency signal, called the carrier. Following the oscillator is the buffer-amplifier stage which amplifies the output of the oscillator and isolates the oscillator from the power amplifier. The final stage is the power amplifier which delivers energy to the antenna. Notice that the output waveshape of the final RF stage does not resemble the input waveshape from the buffer. The RF waveshape has been altered by modulation. This brings us to the modulation or audio section. The microphone converts the sound that is to be transmitted into electrical variations. The weak output of the microphone is fed into an audio amplifier (speech amplifier). The output of the speech amplifier drives an audio power amplifier called a MODULATOR. The modulator injects the audio signals into the RF power amplifier to produce the modulated RF output. The output of the modulated final RF amplifier is fed to the antenna where it is radiated out into space.

10-4 CHECKING THE QUALITY OF A TRANSMITTER'S EMISSIONS

It is important for an amateur to occasionally check the quality of his transmitter's emission. There are many ways that this can be accomplished. The simplest way is to listen to his sending on his receiver. While simple, this method has a serious drawback. A strong signal will overload a receiver's front end and may block it to a point where it is inoperative. To overcome this, we can remove the receiving antenna and reduce the receiver's gain control.

Another way to check the quality of a CW transmitter's output is to use a separate monitor. In this way, we do not disturb the receiver.

A very practical way of checking a transmitter is to get an objective report from a ham located some distance away from the transmitter. Or better still, allow an amateur to operate your transmitter and listen to it on a receiver located some distance from the transmitter.

In addition to checking for the percentage of modulation in a radiotelephone transmitter, the radio operator should check for the following characteristics:

(1) The audio present in the output of the transmitter should be undistorted.

(2) The signal should not "splatter" or cause interference to neighboring frequencies.

(3) The signal should be free of background noise.

10-5 TYPES OF TRANSMISSION

There are various types of transmissions that are used in radio communication. The two methods that we have studied are:

1. Continuous wave transmission in which the carrier is keyed according to the telegraph code.
2. Amplitude modulation transmission in which the carrier is amplitude modulated with audio.

The Federal Communications Commission classifies the various types of emissions by letters and numbers. Some of the more popular types are:

TYPE AØ is the steady unmodulated emission of a CW transmitter. It is used only in special cases, such as radio beacon stations.

TYPE A1, TELEGRAPHY, is the keyed emission of a CW transmitter. It can only be picked up by special receivers.

TYPE A2, MODULATED TELEGRAPHY, is the keyed emission of a transmitter whose carrier is modulated by a pure audio note. It can be picked up by an ordinary receiver.

TYPE A3 RADIOTELEPHONY, is the emission of a transmitter whose carrier is amplitude modulated by voice, music, etc. An example of this type is the familiar broadcast transmitter.

PRACTICE QUESTIONS — LESSON 10

1 Type A1 transmission refers to:
a. AM transmission
b. modulated telegraphy
c. unmodulated telegraphy
d. facsimile

2 The symbol for radiotelephony is:
a. A1
b. A2
c. A3
d. A4

3 A monitor is used to:
a. check the quality of a receiver
b. check the quality of a transmitted signal
c. test an antenna
d. load a transmitter

4 An AM transmitter consists of a CW transmitter and:
a. sideband generator
b. a frequency converter
c. an audio amplifier system
d. an IF stage

5 The frequency of an RF signal is called:
a. AM
b. carrier frequency
c. radio wave
d. A1

6 A CW transmitter does not contain a/an:
a. key.
b. oscillator.
c. modulator.
d. R.F. amplifier.

SECTION III — LESSON 11
ANTENNAS, TRANSMISSION LINES AND FREQUENCY MEASUREMENT

11-1 ANTENNA RADIATION

Once the RF signal has been generated in the transmitter, some means must be provided for radiating this RF energy into space. The transmitting antenna provides the link or impedance matching device between the output stage of the transmitter and space. This energy, in the form of an electromagnetic field, travels through space and cuts across a receiving antenna, inducing a voltage in it. If the receiver is tuned to the same frequency as the transmitter, the signal will be received and heard.

11-2 PROPAGATION OF RADIO WAVES

The radio wave that leaves a transmitter takes two general paths. One path is along the surface of the earth and is called the GROUND WAVE. The other path is towards the sky and is called the SKY WAVE.

In traveling along the surface of the earth, the GROUND WAVE gradually loses its strength until it is completely diminished. On the other hand, the sky wave can travel for thousands of miles.

Some distance above the earth, the sky wave strikes a gaseous mass called the IONOSPHERE. Here, the wave is reflected back to the earth (See Fig. 11-1). If a receiver is located between the end of the ground wave and the point where the sky wave returns to the earth, it will not pick up the transmitted signal. The area between the ground wave zone and the point where the sky wave hits the earth is called the SKIP ZONE. After the wave strikes the earth, it may again be reflected up to the ionosphere and back to the earth. In this way, a signal can travel all around the world.

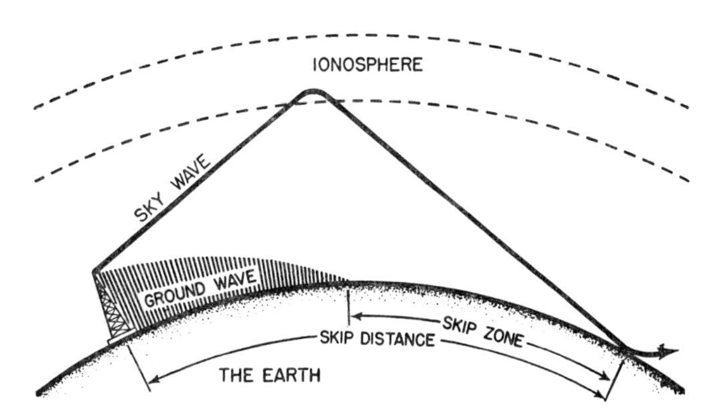

Fig. 11-1. Propagation of radio waves.

Frequencies above 50 MHz, generally do not reflect from the ionosphere. They penetrate the ionosphere and never return to earth.

11-3 ATMOSPHERIC EFFECTS ON PROPAGATION

The atmosphere just above the earth is subject to a wide range of temperature variations that directly affect the propagation of radio energy. The primary and most consistant cause of temperature variations results from the change in seasons. In addition, the changes in temperature resulting from cooler nights compared to sunlit days also affects the ability of the atmosphere to conduct and reflect radio waves.

11-4 THE IONOSPHERE

As stated above, the ionosphere is a gaseous region. It is in the upper atmosphere and extends approximately 30 to 300 miles above the earth. It consists of several layers of ionized particles. The ionization is caused by the sun's ultra violet rays and cosmic rays.

When a radio wave strikes the ionosphere, it is reflected or REFRACTED back to the earth some distance away from the transmitter. The ability of the ionosphere to return a radio wave back to earth depends upon the ion density, the frequency of the signal, the angle of radiation and other factors.

The layers which form the ionosphere undergo considerable variations in altitude, density and thickness, due to the varying degrees of solar or sunspot activity. Every 11 years, the concentration of solar radiation (sunspot activity) into the earth's atmosphere reaches a peak. We refer to this as the SUNSPOT CYCLE. During periods of maximum sunspot activity, the ionized layers are more dense and occur at higher altitudes. They allow for communication over greater distances. The opposite happens during minimum sunspot activity. See Figure 11-2.

In addition to variations in sunspot activity, propagation of radio waves through the ionosphere is affected by a phenomenon referred to as ionospheric storms. Ionospheric storms are caused by eruptions on the surface of the sun which cause disturbances in the atmosphere. Ionospheric storms almost invariably result in reduced or disrupted radio communication for periods ranging from several hours to several days.

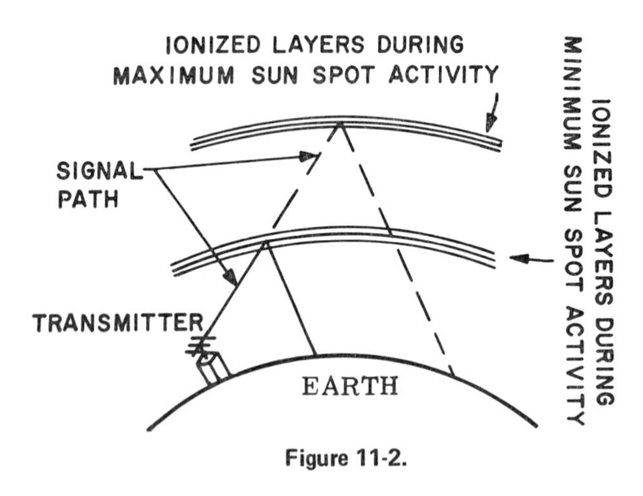

Figure 11-2.

11-5 NOMENCLATURE OF FREQUENCIES

The Federal Communications Commission has set up a system whereby the frequency spectrum is broken down into ranges and classified by name. This system has been universally adopted and is used by everybody in the electronics field. The frequency subdivisions and their ranges are as follows:

Frequency subdivision	Frequency range
VLF (very low frequency)	Below 30 kHz
LF (low frequency)	30 to 300 kHz
MF (medium frequency)	300 to 3,000 kHz
HF (high frequency)	3,000 to 30,000 kHz
VHF (very high frequency)	30,000 kHz to 300 MHz
UHF (ultra high frequency)	300 MHz to 3,000 MHz
SHF (super high frequency)	3,000 MHz to 30,000 MHz
EHF (extreme high frequency)	30,000 MHz to 300,000 MHz

11-6 PROPAGATION IN THE HAM BANDS

Following is a summary of radio wave propagation. Since there are many factors and variables that influence radio wave propagation, the following chart can only serve as an average guide. The first column lists the frequencies, the second column lists the propagation characteristics of the ground wave and the third column lists the propagation characteristics of the sky wave.

FREQUENCY	GROUND WAVE	SKY WAVE
Low Frequency (50 kHz - 500 kHz)	Communications are possible up to 1,000 or more miles, day or night, with high power transmitters.	Communications are generally reliable, day or night, over most of the band. They are slightly better at night. At the upper end of the band, the sky wave is useful only at night. Depending upon the time of day, the seasons and other factors, useful reception can be had for distances up to 8,000 miles.
Broadcast Frequencies (500 - 1600 kHz)	Reception can be had up to 50 to 100 miles, day or night.	There is no sky wave reception in the day time. At night, reception can be had up to 3,000 miles.
160 Meter Amateur Band (1.8 - 2.0 MHz)	Communications are reliable up to approximately 30 miles, day or night.	There is very little sky wave reception during the day. During the night, especially during winter season disturbances, communications can be had up to 2,500 miles.
80 Meter Amateur Band (3.5 - 4.0 MHz)	Communications are good only up to about 20 miles.	During the day time, useful communications are possible up to about 150 to 250 miles. At night time, communication is possible up to 2,000 to 3,000 miles.
40 Meter Amateur Band (7.0 - 7.3 MHz)	Communications are good only up to about 20 miles.	During the day, useful communication can be had up to 750 miles. At night, communication is possible up to 10,000 miles.
20 Meter Amateur Band	Communications are good only up	During minimum sunspot activity, there is almost world wide com-

FREQUENCY	GROUND WAVE	SKY WAVE
(14.0 - 14.35 MHz)	to about 20 miles.	munications during daylight hours and almost no communications at night. During medium sunspot activity, world wide reception is possible during the daylight hours and during the early evening hours. At the peak of sunspot activity, excellent reception is possible for almost 24 hours.
15 Meter Amateur Band (21.0 - 21.450 MHz)	Communications are good only up to about 20 miles.	World wide communications during day and night hours is possible during maximum sunspot activity. During minimum sunspot activity, there is no night time communication and some daylight communication.
10 Meter Amateur Band (28.0 - 29.7 MHz)	Communications are good only up to about 20 miles.	During maximum sunspot activity, excellent communication can be had during daylight hours and early evening hours. There is generally little communication at night. During minimum sunspot activity, the band is "dead" except for local communication.
Ultra High Frequency (30 - 300 MHz)	There is little or no ground wave propagation. The only communication is through the direct, line of sight wave, from the transmitter antenna to the receiver antenna.	There is generally very little reflection from the ionosphere, day or night. Occasionally, there is some sporadic reception for short periods of time in limited localties.

11-7 PRINCIPLES OF RADIATION

The currents flowing in the antenna, due to the excitation from the transmitter, set up magnetic and electrostatic fields which are pushed out from the antenna and fly off into space in all directions. The two fields, moving through space as an electromagnetic wave, contain the carrier and sideband energy and, as such, have quite definite characteristics. These characteristics are:

1. The wave has a very definite frequency which is equal to the carrier frequency of the transmitter.

2. The wave travels through space at a constant velocity, regardless of the frequency at which it is being transmitted. This velocity is 186,000 miles per second or 3×10^8 meters per second.

$$3 \times 10^8 = 300{,}000{,}000$$

3. The wave has a certain wave length which is defined as the distance between adjacent peaks, or the distance the wave travels through space during one cycle of the antenna current. The wave length is measured in meters and is given the symbol "L".

L = wavelength in meters.

4. An equation which ties together wavelength, frequency and velocity of an electromagnetic wave is:

$$(1) \qquad L = \frac{300{,}000{,}000}{F}$$

where F is the frequency of the wave in Hertz.
L is the wavelength in meters.

300,000,000 is the velocity of the electromagnetic wave in free space, which is constant.
If the frequency is in kilohertz, the formula becomes:

$$(2) \qquad L = \frac{300{,}000}{F}$$

where L is in meters and F is in kHz.

If we wish to solve for frequency, the formula becomes:

$$(3) \qquad F = \frac{300{,}000}{L}$$

where L is in meters and F is in kHz.

A few examples will clarify the use of the above formulas.
(a) Find the wavelength of the distress frequency, 500 kHz.
SOLUTION: Use formula No. 2.

$$L\ (\text{meters}) = \frac{300{,}000}{F(\text{kHz})} = \frac{300{,}000}{500} = 600 \text{ meters}$$

(b) Find the frequency of the signal whose wavelength is 300 meters.
SOLUTION: Use formula No. 3.

$$F\ (\text{kHz}) = \frac{300{,}000}{L\ (\text{meters})} = \frac{300{,}000}{300} = 1000 \text{ kHz.}$$

The wavelength for very high frequencies can be given in centimeters. A centimeter is one hundredth of a meter.

11-8 FUNDAMENTAL ANTENNA CONSIDERATIONS

Figure 11-3 shows an antenna or wire connected to an RF source. The alternating current travels out from Point A and along the wire until it reaches point B. Since point B is free, the wave cannot continue farther and bounces back, or is reflected, from this point. The distance an RF wave travels during the period of one cycle is known as the wavelength. If the wave is to travel exactly the length of the wire and back, during the period of one cycle, it is evident that the wire must be equal in length to one half the wavelength of the voltage being applied. The wire is then said to be resonant to the frequency of the applied voltage. During the negative alternation of the RF generator, electrons will move along the wire away from point A towards point B. The electrons are stopped and accumulate at point B, which represents a high voltage point. During the positive alternation of the

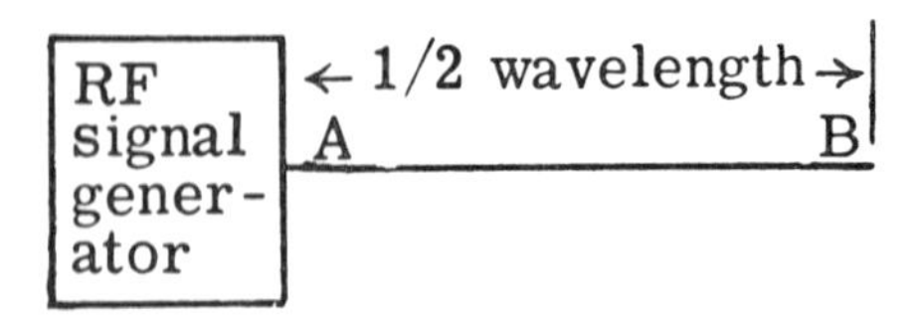

Figure 11-3. Half-wave antenna.

RF power source, electrons move away from point B and crowd together at point A, which also represents a high voltage point. In the center of the antenna there is, at all times, a maximum movement of electrons causing a high current or a low voltage point. Therefore, very little voltage will appear at the center of the antenna and no current will flow at the ends. Figure 11-4 illustrates the voltage and current distribution on a fundamental half wave antenna. This representation of a voltage and current distribution is known as a standing wave pattern. The points of minimum current and minimum voltage are known as current and voltage nodes respectively. An antenna is said to be resonant when there exist standing waves of voltage and current along its length. Since the waves traveling back and forth in the antenna reinforce each other, a maximum radiation of electro-magnetic waves into space results. When there is no resonance (no standing waves), the

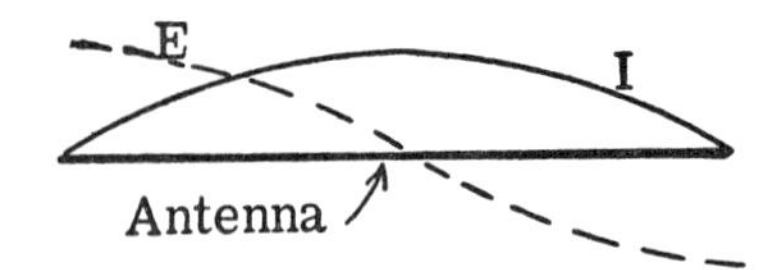

Figure 11-4. Distribution of voltage and current on a half-wave antenna.

waves tend to cancel each other, thus dissipating their energies in the form of heat loss, rather than utilizing them to radiate the radio waves. Therefore, a resonant antenna connected to an RF generator can dissipate power because most of the energy leaves the antenna in the form of radiation.

11-9 ANTENNA IMPEDANCE

Since voltage and current vary along the length of the antenna, a definite impedance value must be associated with each point along the antenna. The impedance varies according to the relative crowding of the electrons as the ends are approached. The impedance existing at any point is simply the voltage at that point divided by the current at that point. Thus, the lowest impedance occurs where the current is highest; and the highest impedance occurs where the current is lowest. When we speak about antenna impedance, we usually mean the impedance at the point where the transmission line feeds the antenna. The impedance at this point is referred to as the INPUT IMPEDANCE.

11-10 PRACTICAL TRANSMITTING ANTENNAS

Most practical transmitting antennas come under one of two classifications, Hertz antennas or Marconi antennas. A Hertz antenna is operated some distance above the ground and may be either vertical or horizontal. A

Marconi antenna operates with one end grounded (usually through the output of the transmitter or the coupling coil at the end of the feed line). Hertz antennas are generally used at higher frequencies, above about two megacycles, while Marconi antennas are used at lower frequencies.

11-11 THE HERTZ (DIPOLE) ANTENNA

A Hertz antenna is any length of wire far enough from ground so that it will not be influenced by grounded objects. Therefore, its physical length will directly determine the wavelength to which it will tune. A short length antenna will be resonant to a short wavelength or a high frequency; a long length antenna will be resonant to a long wavelength or low frequency. Therefore, the resonant frequency of a Hertz antenna can be changed by

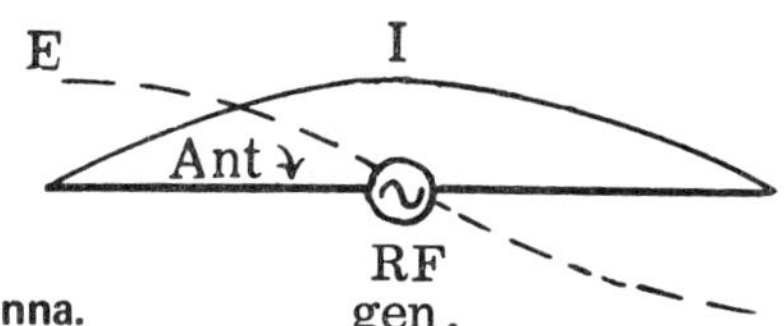

Figure 11-5. Center-fed Hertz antenna.

varying its physical length. This is true because an antenna acts like a resonant circuit. Figure 11-5 illustrates a center-fed, half-wave Hertz antenna. The physical length of the antenna is one-half of the wavelength of the signal that it is radiating. This type of antenna is also referred to as a "dipole" (or "half-wave dipole") antenna.

Since the center of a half-wave antenna is a high current point (see Paragraph 11-8), we can say that this antenna is current-fed by the transmitter. The impedance at the center of this Hertz antenna is about 73 ohms. The impedance rises uniformly towards each end of the antenna where it is about 2400 ohms.

In order to find the approximate electrical length of a half-wave antenna, we can use Formula 2 on Page 84 and divide the result by 2. However, due to end effects, diameter variation and other factors, the actual antenna should be about 5% less than the electrical length computed from the basic formula. A more accurate and practical formula to use in finding the physical length of an antenna, if the frequency is known, is:

$$L = \frac{468}{f}$$

Where: L is the length in feet and f is the frequency in MHz.

FOR EXAMPLE:

QUESTION: Find the length of a half-wave antenna for the middle of the 40 meter Novice ham band.

ANSWER: From the Rules and Regulations (Lesson 14), we find that the middle of the 40 meter ham band is 7.125 MHz. We then substitute in the formula.

$$L = \frac{468}{7.125} = 65.68 \text{ feet.}$$

11-11A THE QUARTER WAVE MARCONI ANTENNA

If one half of a half-wave Hertz antenna is replaced by a conducting plane, the remaining quarter-wave will continue to radiate properly. This assumes that the conducting plane is large and acts as a good ground. The ground actually makes up for the other half of the antenna's electrical length. In other words, the antenna proper provides one-quarter wavelength and the ground supplies the additional one-quarter wavelength. The Marconi antenna is a practical example of this type of vertical antenna. Since the Marconi antenna is only one-quarter wavelength, it is one half the physical length of a Hertz antenna and therefore, is more practical as a vertical antenna for mobile operation.

11-12 HARMONIC OPERATION OF A DIPOLE ANTENNA

The most efficient antenna system is one where the antenna is cut for a certain frequency and is used for that one frequency or band of frequencies. If an amateur wishes to operate efficiently on several bands, he must use several antennas. This would take up a considerable amount of space and be quite costly. There is an alternative to this. An antenna which is a half wavelength long at one particular frequency is two half-wavelengths long at twice the frequency and three one half-wavelengths long at three times the basic frequency. Fortunately, this antenna will act as a reasonably good antenna for all three frequencies. The rule is that, as long as the length of the antenna is an integral multiple of a half wavelength, it will act as an antenna. Since the amateur band frequencies are harmonically related to each other, it is easy to see how a single length of wire can serve as an antenna for several amateur bands.

For instance, a half wave antenna for the 7 MHz (40 Meter) band, is approximately 20 meters long. This represents two one-half wave lengths on 14 MHz (20 meters) and four one-half wave lengths on 28 MHz (10 meters). Thus, the 20 meter length of wire can act as an antenna on the 7 MHz, 14 MHz, and 28 MHz ham bands.

While multiband antennas are practical, they do have drawbacks. For one thing, they tend to radiate harmonics with its subsequent interference. Their efficiency and directivity are not as good as a single band antenna. It is also difficult to achieve a low SWR on all frequencies with a multiband antenna.

One way to reduce the emission of harmonics is to use a "TRANSMATCH". The transmatch is inserted between the transmitter and the antenna transmission line. It consists of one or more tuned circuits (coils and capacitors) that selects the desired signal leaving the transmitter and rejects all others. Another important function of the transmatch is to match the impedance of the transmitter output to the transmission line. Other advantages of the transmatch are: It permits a transmitter to feed into almost any type of antenna, it permits a transmitter with an unbalanced output to feed a balanced transmission line, and vice versa, and it permits maximum RF to be transferred from the transmitter to the antenna system. The transmatch receives its RF power from the transmitter and is not connected to the AC power line. Other ways of minimizing harmonic radiation from antennas are:

(1) Use a low-pass filter between the transmitter and the transmission line. This is described in Lesson 9.

(2) Make certain that all antenna joints are mechanically sound and free of corrosion. Corroded joints can act as rectifiers for harmonics and cause their radiation.

11-13 DUMMY ANTENNA

When a transmitter is being tuned up for optimum operation, the antenna should be coupled to the final stage in order to insure correct settings of plate voltage and current (since the antenna is the load of the final stage).

Coupling an antenna during the adjustment period is forbidden by law, since radiation will result which may cause interference. Most stations get around this difficulty by utilizing a dummy antenna which is nothing more than a resistive load of the correct power dissipation coupled to the tank coil of the final output tube in the same manner as the antenna is later to be coupled. An incandescent bulb of the proper wattage can readily be used. The brilliance of the lamp will give a rough idea of the transmitter power output. The peak brilliance of the lamp will indicate to the operator when the transmitter is tuned for maximum power output.

11-14 TRANSMISSION LINES

In practically all transmitter installations, the antenna is located some distance from the transmitter. It may be 10 feet or it may be 200 feet. In all cases, some means must be used to carry the RF energy from the output of the transmitter to the antenna. The lines that are used to carry this energy are called transmission lines.

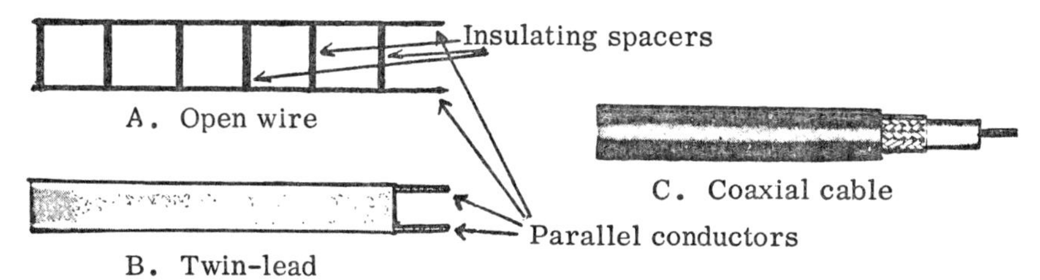

Figure 11-6. Transmission lines.

There are two basic types of transmission lines. One is called parallel line; the other is called coaxial cable. The parallel-line is further subdivided into two types. One is called the "open-wire" type. The other is the "twin-lead" type. Open-wire transmission line consists of two conductors in parallel separated by insulating spacers. The twin-lead type consists of two parallel conductors separated by flexible insulation. This type is commonly used in television installations. See Figure 11-6.

Coaxial cable consists of a conductor surrounded by a round flexible insulator. There is a concentric metallic covering made of flexible wire braid around this insulator. Surrounding this is a weatherproof vinylite sheath. The outer conductor acts as a shield, preventing spurious and harmonic radiation from the transmission line.

The transmission line should carry the RF energy from the transmitter to the antenna with minimum losses. In order for this to occur, the impedance of the transmission line must be equal to the impedance of the transmitter's output and to the impedance of the antenna. Matching the impedances also cuts down on the radiation of spurious and harmonic signals.

The characteristic impedance of a parallel wire air insulated transmission line can be found with the aid of the following formula:

$$Z_0 = 276 \log \frac{b}{a}$$

Where: Z_0 is the characteristic impedance of the line
b is the center to center spacing between conductors
a is the radius of the conductors

The characteristic impedance of a coaxial transmission line is given by the following formula:

$$Z_o = 138 \log \frac{b}{a}$$

Where: b is the inside diameter of the outer conductor
a is the outside diameter of the inner conductor

11-15 STANDING WAVE RATIO

It was seen in paragraph 11-8 of this lesson that a standing wave pattern occurs along the length of an antenna. These standing waves result in maximum radiation of the signal and they are desirable in an antenna. However, standing waves are undesirable on a transmission line. We do not want the transmission line to radiate; we only want it to transfer energy. We therefore do not want standing waves on a transmission line.

Standing waves occur on a transmission line when some of the power that travels along the line to the antenna is reflected back to the transmitter instead of going into the antenna. This happens when the impedance of the line is not equal to the impedances of the transmitter and antenna.

The standing waves on a line give us maximum and minimum voltage and current points along the line. The ratio of the maximum current to minimum current or maximum voltage to minimum voltage, is called the STANDING WAVE RATIO (SWR). Expressed as a formula, it is:

$$SWR = \frac{I_{max.}}{I_{min.}} \text{ or } \frac{E_{max.}}{E_{min.}}$$

Where: SWR is the Standing Wave Ratio
$I_{max.}$ and $E_{max.}$ are the maximum current and voltage points on the line.
$I_{min.}$ and $E_{min.}$ are the minimum current and voltage points on the line.

In a transmission line, we want a minimum standing wave ratio. In other words, we want the current and voltage to be the same along the entire line.

One way to determine the standing wave ratio is to measure the voltage of the RF energy traveling to the antenna, and to compare it to the voltage of the RF energy being reflected back from the antenna. We can use these two voltages in the following formula to find the SWR:

$$SWR = \frac{E_F + E_R}{E_F - E_R}$$

where: E_F is the forward or incident voltage and E_R is the reflected voltage.

The ideal SWR occurs when the reflected voltage is zero. We can see from the formula that when E_R is zero, the SWR is 1/1 or 1 to 1. As the reflected voltage rises, the SWR increases to 2 to 1, 3 to 1, etc.

An SWR in excess of 3 to 1 is considered high and should be avoided. The ideal SWR is 1 to 1. 2 to 1 is acceptable.

A high SWR is generally caused by an improper impedance match between the transmission line and the antenna. The impedance of the antenna at the point where it is fed should be equal to the impedance of the transmission line.

An instrument, called an SWR bridge, or REFLECTOMETER, reads the forward and reflected voltages, and shows the SWR on the face of its meter.

11-16 ANTENNA SUPPORT, INSTALLATION, MAINTENANCE AND SAFETY

Since most amateur antennas are designed for installation out of doors, it is essential that the location, supports, and accessability for adjustment and maintenance be considered prior to the actual installation. The location of the antenna is usually limited by the physical size of the antenna. Horizontal antennas such as a dipole for the higher wavelengths are particularly long and are therefore limited in their location requirements.

Vertical antennas present less of a problem in that they require very small amounts of horizontal space. However, since they are usually supported at the base end only, they are more subject to the effects of wind than are horizontal wire type antennas.

In addition to location, the most critical factor to be considered in the installation of an antenna is the means by which the antenna is to be supported. A poorly supported antenna will present more problems than if proper support was provided to begin with. Horizontal and V-type wire antennas should be supported by strong wire or guy ropes with high strength insulators. Always try to install the antenna between two strong supporting members such as a house at one end and a tree or pole at the other end.

Since vertical antennas are supported at the base only, it is essential that particular care be taken to provide a good base mount and, if possible, use insulated guy wires or ropes to support the antenna about one-third of the way up from the base.

The antenna should be mounted in a position that will allow for easy maintenance such as tuning (pruning the length), resecuring the supports, and access to transmission line terminals.

If you are installing a large antenna or tower and do not have the experience or equipment, you should hire a professional insured firm to do the work.

If you are going to do the work yourself, try to get help from other experienced hams. Use a safety belt in climbing a tower or other high installations. Stay clear of electrical wires. In handling the antenna, make sure that the parts of the antenna do not touch any overhead electric wires.

Use the following construction procedures in the installation to make sure that it is not necessary to repair the antenna on a constant basis.

(1) Use large strong insulators.

(2) Use heavy conductor wires, stranded, if possible.

(3) Use non-corrosive hardware.

(4) The antenna system should be able to withstand the strongest winds or storms that may occur in the area.

Use a good ground for the antenna system. A good ground can be made by burying an eight foot metal rod into the earth.

In general, use good common "safety-sense" in the installation and maintenance of the antenna system. Remember that ham radio is a hobby and that it doesn't pay to take chances with life or limb in the pursuit of a hobby.

11-17 PROTECTING STATION EQUIPMENT FROM STATIC CHARGE AND LIGHTNING DISCHARGE ON ANTENNA

During an electric storm, an antenna may be hit by lightning which can damage the station equipment. A means must be provided to safely bypass the lightning discharge to ground. What is usually done is to connect spark gaps of large current carrying capacity between the antenna and ground. The spark gap will provide an effective by-pass for the lightning surge. If an antenna is capacity coupled to the output of the transmitter, static charges of high potential may build up on the antenna because there is no direct leakage path to ground. In this case, static drain coils having a high resistance at the radiating frequency are connected from the base of the antenna to ground. They serve as the discharge path for any static charge on the antenna. An antenna grounding switch can also be used to discharge the atmospheric electricity that accumulates on the antenna system. During transmission or reception, the switch should be open.

11-18 GROUND

Cold water pipes are effective grounds since they make excellent contact with the earth. The ground terminals of the radio station and the antenna system should be connected by means of a heavy conductor to a cold water pipe. If there is no cold water pipe system, an effective ground can be made by driving a copper rod of at least 6 to 8 feet into the earth. We call this a "ground rod". Even where a cold water system is used, it would help to use a ground rod in addition to the cold water pipe for ground. The ground rod could be located close to the antenna tower and connected to the ground of the antenna system. A heavy conductor could then be used to connect the ground rod to the cold water pipe and the ground terminals of the radio station. The circuit reference to a physical ground such as a ground rod or chassis is referred to as GROUND POTENTIAL.

11-19 THE MEASUREMENT AND DETERMINATION OF FREQUENCY

The FCC states that a station must maintain its exact operating frequency so that stations do not interfere with each other. Therefore, one of the most important duties of a radio operator is to keep his station exactly on frequency. The FCC assigns definite frequencies or bands of frequencies to the various transmitting services. Operating "off frequency" represents a serious offense and must be avoided. Therefore it is very important to be able to measure the frequency of a transmitter. The instrument which is used to measure frequency is called a FREQUENCY METER. We will discuss several different types of frequency meters.

(1) ABSORPTION-TYPE WAVEMETER. An absorption-type wavemeter consists of a coil and capacitor in parallel and an indicator. The indicator can be a flashlight bulb or a sensitive microammeter. The wavemeter coil is brought near the output tuning circuit of the transmitter. The capacitor of the wavemeter is varied until the indicator reads or glows. At this point, the resonant frequency of the wavementer circuit is the same as the transmitter's output frequency. The wavemeter can be calibrated to show the frequency. The absorption-type wavemeter is simple to use, but is not highly accurate.

(2) HETERODYNE FREQUENCY METER. When two signals are mixed together, two new signals are produced in addition to the two original signals. The two new signals have frequencies that are equal to the sum and difference of the two original signals. If the two original signals are equal in frequency, one of the two new signals will have a frequency of zero. This principle is made use of in the Heterodyne Frequency Meter.

The Heterodyne Frequency Meter consists of a stable variable frequency oscillator, a detector, and possibly, a crystal-controlled oscillator. The incoming signal from the transmitter and the signal of the variable frequency oscillator are fed to the detector where they mix together. The frequency of the variable frequency oscillator is varied until a "zero-beat" indication tells us that it is the same as the incoming transmitter frequency. By reading the dial of the meter, we can tell the frequency of the transmitter. The crystal-controlled oscillator is used to accurately calibrate the meter. The Heterodyne Frequency Meter is a highly accurate method of measuring one's frequency.

(3) FREQUENCY MARKER. The simplest method of checking one's transmitter frequency is to tune it in on his receiver, being careful not to overload the receiver. The operator then looks at his dial to see what his frequency is. However, there is no assurance that the receiver dial is accurate. In order to correct the receiver calibration, we use a frequency marker. The frequency marker is a highly stable oscillator that is rich in harmonics. A popular frequency marker uses a 100 kHz crystal and produces signals every 100 kHz. With these marker frequencies, it is a simple matter to adjust the receiver so that the dial is accurate.

Many receivers and transceivers have a frequency marker built into them.

PRACTICE QUESTIONS — LESSON 11

1 An advantage of a Transmatch is not:
- a. it reduces audio distortion
- b. it brings about a good impedance match
- c. it enables the transmitter to be used with several types of transmission lines
- d. it reduces television interference

2 An important reason for using a dummy antenna is:
- a. to minimize interference
- b. it is the only way an FM transmitter can be tested
- c. to measure the output power
- d. to reduce splatter

3 If the voltage and current are the same on all points of a transmission line, the SWR is:
a. 1 to 1 b. 2 to 1 c. 1 to 2 d. .636 to 1

4 On which of the following ham bands is there maximum reception during daylight hours and during minimum sunspot activity?
a. 160 meters b. 40 meters c. 20 meters d. 10 meters

5 Ionization of the upper atmosphere is primarily dependent upon radiation from the:
a. outer space b. sun c. moon d. earth

6 What is the approximate length of a half-wave antenna that is to transmit a frequency of 4850 kHz?
a. 110 feet b. 96 feet c. 86 feet d. 79 feet

7 A characteristic of a multiband antenna is:
a. low SWR b. high efficiency c. difficult to feed d. high gain

8 The sky wave of a transmitter is refracted back to earth by the:
a. sun spots b. ionosphere c. antenna reflector d. SWR

9 A transmitter is protected from damage from lightning by:
a. connecting spark gaps from the antenna base to ground
b. shorting the transmission line.
c. place a capacitor in series with the antenna.
d. using coaxial cable.

10 Which of the following is not used as a transmission line in an amateur transmitting antenna system?
a. TV twin lead
b. RF waveguide line
c. parallel conductors separated by insulators
d. coaxial cable

11 What is the exact wavelength of a 21,180 kHz signal?
a. 51.18 meters b. .0706 meters c. 14.1 meters d. 70.6 meters

12 Which of the following is used to determine if a transmitter is operating within a specific band?
a. watt-hour meter
b. a galvanometer
c. a frequency meter
d. a band meter

13 The area between the ground wave zone and the point where the sky wave returns to earth is called the:
a. skip zone. b. ionosphere. c. sun spot. d. SWR zone.

14 An advantage of coaxial cable over other types of transmission line is:
a. it is inexpensive.
b. it has the lowest losses.
c. it will not radiate spurious signals.
d. it matches all types of antennas.

SECTION III – LESSON 12
THE RADIO RECEIVER

12-1 FUNCTIONS OF A RADIO RECEIVER

Up to this point, we have covered the principles of transmitters in detail. To complete the picture, we must consider the problem of the reception of radio waves by the radio receiver.

The radio receiver must be able to perform the following functions:

1. pick up radio frequency signals radiated by transmitters.

2. tune to one desired signal and reject the remaining signals.

3. amplify the desired radio frequency signal using radio frequency amplifiers.

4. detect or demodulate the desired signal (separate the audio intelligence from the radio frequency carrier).

5. amplify the detected audio signal and drive a speaker or headphones with it.

12-2 THE ONE TUBE RECEIVER

Figure 12-1 illustrates a block diagram of a one tube radio receiver.

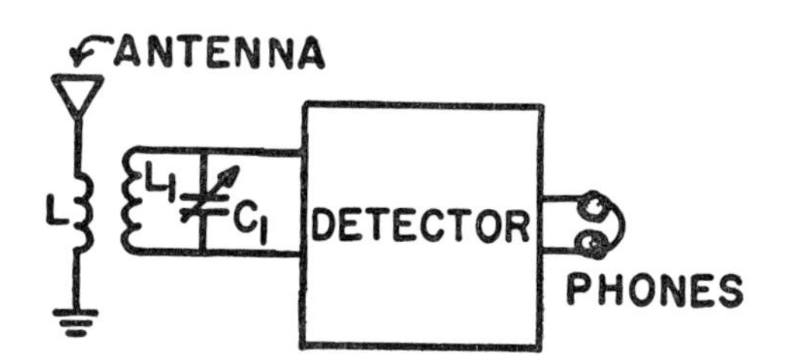

Figure 12-1. A simple one-tube receiver.

The antenna picks up any radiated signals that may be present in its vicinity and couples them to the tuned circuit L_1-C_1. The function of the tuned circuit is to select the station that is desired to be heard and at the same time, to reject the unwanted signals. This is accomplished by varying capacitor C_1, until the tuned circuit is resonant to the desired frequency. For a review of the theory of a tuned circuit, refer to Lesson, 3 Paragraph 9 on resonance. Once the signal has been selected, it is necessary to extract the audio-frequency intelligence from the radio-frequency carrier. This is done by the detector, or demodulator, stage. A simple diode is generally used in the detector stage. The audio is then applied to the headphones.

12-3 THE TUNED RADIO-FREQUENCY RECEIVER

A radio signal diminishes in strength at a very rapid rate after it leaves the transmitting antenna. Therefore, it is seldom possible to use a detector circuit alone to obtain any useful output from the few microvolts of signal available at the receiving antenna. To remedy this, it is desirable to amplify the RF signal before it is detected. This is done by the use of an RF amplifier. The RF amplifier, like the detector, is provided with one or more tuned circuits so that it amplifies only the desired signal. The addition of an RF amplifier to the receiver gives not only greater sensitivity (ability to receive weak signals), but also greater selectivity (ability to separate signals). Audio amplifier stages usually follow the detector to amplify the audio signals before they are applied to the reproducer. The complete receiver, consisting of radio-frequency amplifiers, detector and audio amplifiers, is called the tuned radio frequency receiver or, as it is more commonly called, the TRF receiver.

A block diagram of a TRF receiver, showing the signal passing through the receiver, is illustrated in Figure 12-2.

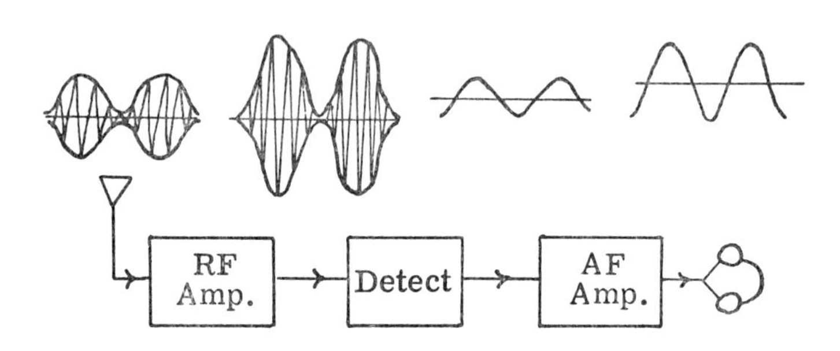

Figure 12-2. Block diagram of a TRF receiver.

12-4 CAPABILITIES OF A TRF RECEIVER

Although the TRF receiver will give satisfactory results when covering the medium frequency band, such as the broadcast band, it has several disadvantages which makes it impractical for use in high frequency and multiband receivers. The chief disadvantage of the TRF receiver is that its selectivity does not remain constant over its tuning range. When a tuned circuit is made variable, as it must be in a TRF receiver, selectivity will decrease as the receiver is tuned to the high end of the band. If it were possible to design a receiver in which the selective circuits were fixed tuned, these circuits could very easily be designed for high gain and selectivity for the particular frequency at which they are to operate. This desirable effect is accomplished in the SUPERHETERODYNE RECEIVER.

12-5 THEORY OF SUPERHETERODYNE ACTION

The important difference between the TRF receiver and the superheterodyne receiver is that in the TRF, the RF signal is amplified at the frequency of the signal, while in the superheterodyne receiver, the RF signal is amplified at a new, lower, fixed frequency called the INTERMEDIATE FREQUENCY. The intermediate frequency (IF), though much lower in frequency than the original signal, retains all the modulation characteristics of the original signal. By amplifying this lower fixed frequency, it is possible to use circuits which are more selective and capable of greater amplification than the circuits used in TRF receivers.

12-6 THE SUPERHETERODYNE RECEIVER

The block diagram of a typical superheterodyne receiver is illustrated in Figure 12-3. The received modulated RF signal is first passed through an RF amplifier. A locally generated unmodulated RF signal is then mixed with the carrier frequency in the mixer stage. The mixer is called a converter and sometimes a first detector. This mixing action produces two new modulated RF signals in the output of the mixer, the sum and the difference, in addition

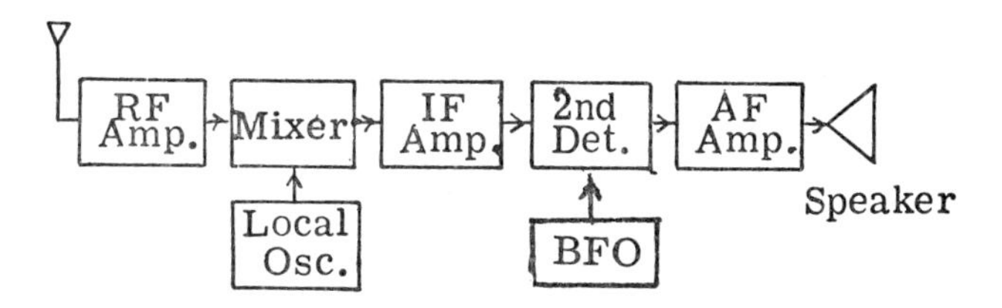

Figure 12-3. Block diagram of a superheterodyne receiver.

to the original signals. It is the difference frequency in which we are interested. The difference frequency is called the INTERMEDIATE FREQUENCY or I.F. A fixed tuned circuit in the plate of the mixer will reject all frequencies except the IF frequency to which it is tuned. This new IF frequency contains all the modulation characteristics of the original signal, but it is much lower in frequency. The intermediate frequency is usually set at some definite value. The frequency of the local oscillator must differ from that of the signal being received by an amount equal to the intermediate frequency. Thus, as the RF amplifier of the receiver is tuned to signals of various frequencies, the local oscillator must be tuned simultaneously, so that its frequency is always separated from that of the signal by the same amount. For example, if the IF is 450 kiloHertz, a commonly used frequency, and the range of the receiver is from 500 to 1600 kHz., the oscillator would have to operate over a range of either 950 (450 + 500) to 2050 (450 + 1600) kiloHertz or 50 (500 - 450) to 1150 (1600 - 450) kiloHertz. Whether the oscillator frequencies are higher or lower than the signal, the difference is still 450 kHz. The intermediate frequency is then amplified in one or more fixed tuned stages called intermediate-frequency amplifiers, and is then fed into the second detector where it is detected or demodulated. The detected signal is amplified in the AF amplifier and then fed to the headset or loudspeaker. The reason why the detector is called the second detector is because the mixer tube is sometimes called the first detector. Because of this, the superheterodyne is sometimes called a double detector receiver.

12-7 BEAT FREQUENCY OSCILLATOR

The detectors that have been discussed thus far are able to detect only phone signals. They are not able to detect a CW signal. You will recall that a detector extracts audio from an RF carrier. However, a CW signal does not have any audio. It is all RF. If we listen to a CW code signal with the detectors thus far described, all we will hear are thumps and clicks.

The solution to this problem is to introduce a signal at the receiver that will mix or beat with the incoming CW signal to produce an audible signal. To accomplish this, the receiver contains a simple oscillator called a BEAT FREQUENCY OSCILLATOR (BFO). It is connected to the detector. See Figure 12-3. The BFO generates a signal that is approximately 500 to 1000 Hz different from the incoming CW signal. The two signals beat together in the detector and the difference, which is audible to the human ear, is present in the output of the detector.

The amplitude of the Beat Frequency Oscillator signal should be greater than the amplitude of the incoming code signals in order to operate properly.

PRACTICE QUESTIONS — LESSON 12

1 The function of the detector is to:
   a. amplify the signal
   b. demodulate the carrier
   c. excite the speaker
   d. bypass the AF

2 A superheterodyne does not contain a/an:
   a. oscillator
   b. detector
   c. RF power amplifier
   d. audio amplifier

3 The Intermediate Frequency:
   a. is the difference frequency between the carrier-frequency and the local oscillator frequency
   b. is always one-half the local oscillator frequency
   c. depends upon the station tuned in
   d. is twice the carrier frequency

4 Fixed tuned IF stages result in:
   a. high fidelity
   b. variable IF
   c. high selectivity, high gain
   d. constant power output

5 A Tuned Radio Frequency Receiver does not contain a/an:
   a. audio amplifier
   b. RF amplifier
   c. detector
   d. IF amplifier

6 Audio is extracted from the RF carrier by a/an:
   a. tuned circuit
   b. detector
   c. IF amplifier
   d. oscillator

7 A TRF receiver is superior to a superheterodyne receiver because:
   a. it provides more gain.
   b. it provides more selectivity.
   c. a and b are correct.
   d. none of the above.

SECTION III — LESSON 13
OPERATING PRACTICES

13-1 WHAT IT'S ALL ABOUT

In addition to being able to obtain your Novice Class license, the end product of all you have learned is your ability to operate your station. Operation of your station requires familiarity with electronics technology as covered in Lessons 1 through 12 and knowledge of the FCC rules and regulations as covered in Lesson 14. The application of your technical skills within the intent of the rules and regulations will be displayed by the manner in which you operate your station. The use of good operating practices assures the operator an enjoyable hobby; bad operating practices are a nuisance to all other amateurs.

13-2 COURTESY

Courtesy is the most essential of all operating practices and its use by all radio amateurs has made Ham Radio one of the most esteemed hobbies in the world. Courtesy means operating your station with respect for your neighbors, the commercial and private radio frequency users, and your fellow amateurs.

Maintaining your station to eliminate interfering parasitics, harmonics, and spurious signals is a primary function of the radio amateur. The elimination of these interfering signals is not only a requirement in compliance with the FCC regulations, it is also essential as an act of courtesy to your neighbors and commercial agencies that share in the use of the radio spectrum. Interference with television, radio, and phonograph entertainment facilities will strain your relationship with your neighbors and will, therefore, limit your enjoyment of the hobby. In addition, the elimination of interference from your station is an act of respect to those entertainment agencies, such as TV and Radio stations, that share the use of the radio spectrum as a communications medium.

Courtesy to your fellow amateurs means maintaining all communications on a level that will not be offensive, and respecting your fellow amateurs as having the same privileges as yourself. The following general rules should serve as a guide:

a. Limit the length of your CQ (general call to all stations) to allow receiving stations to respond. Generally, sending CQ three times, followed by your call letters sent three times, and terminated with a "K" is the accepted standard for a code transmitted general call.

b. If a CQ is received or responded to at a speed lower than your normal code speed, try to lower your speed to accommodate the other amateur.

c. Remember that there are potentially thousands of listeners for each conversation. Always maintain the content of your transmissions at a level than will not be considered offensive to any listener.

13-3 INTERFERENCE AND CONGESTION IN THE AMATEUR BANDS

An amateur must operate his transmitter in a manner so as not to cause interference to other amateurs. Also, some of the bands are quite congested and a good operator must avoid adding to this congestion. The following operating hints will minimize interference and congestion of the amateur bands:

(1) Before operating on a certain frequency, listen in on the frequency to see if it is being used. If the frequency is in use, wait till the parties have finished using the frequency or move on to a clear channel.

(2) When testing a transmitter, use a dummy antenna. Do not test "on the air". If "on the air" tests have to be made, do them on clear frequencies and during hours when there is little activity. Also, be brief.

(3) Use the minimum amount of power necessary to effect proper communications between the two stations.

(4) Do not overmodulate or splatter. This causes interference to adjacent frequencies. Also, make certain that no spurious or harmonic radiations are emitted from the transmitter.

(5) Use the minimum bandwidth necessary for good speech communications. The audio frequencies above 3 kHz should be eliminated to keep the bandwidth of the signal down. It should be noted that a CW signal has an extremely small bandwidth (100 Hz. or less) and occupies very little space in the frequency spectrum. It therefore causes less interference than phone signals.

(6) Use directional antennas. In this way, the signal is confined to fewer areas.

(7) In making CQ directional calls, give sufficient information to avoid useless answers.

(8) Use a recognized phonetic word list to avoid needless repetition and confusion. Also, use Q signals and other known abbreviations to cut down the length of the transmissions.

(9) Use less congested frequencies to make contacts where a choice is possible. For instance, don't use the 20 meter band to contact a local amateur when both of you have two-meter equipment. 20 meters is very crowded and 2 meters is not.

(10) Keep transmissions as brief as possible. Do not repeat words unless it is necessary. Also, in calling, use short calls with frequent breaks for listening.

13-4 CHOOSING AN OPERATING FREQUENCY

Radio amateurs are not assigned specific operating frequencies. They can operate anywhere they want to, within specified band limits. In other words, an amateur holding a Novice license, can operate on any frequency within the 3700 kHz to 3750 kHz band. The operating frequency that is chosen depends upon many factors. If he wishes to communicate with Amateur Station W2ABC who is in contact with Amateur Station W6XYZ, he sets his station frequency at or near W6XYZ and waits until the two amateurs have finished their contact. He then calls Station W2ABC. The reason he sets his frequency on W6XYZ's frequency is that W2ABC is listening to the frequency that Station W6XYZ is transmitting on. In most instances, two stations in contact with each other are on the same frequency.

If an amateur wishes to communicate with any other amateur or if he wishes to communicate with a particular area, he chooses a clear frequency and calls "CQ" or CQ to a particular area.

If an amateur is operating his station close to the edge of the band, he must make absolutely certain that he is not operating outside of the band.

Lesson 11 discusses the various methods that are used to determine the exact operating frequency.

The band that an amateur uses depends upon the distance to be covered, the time of day, the time of the year, etc. The chart in Lesson 11 lists the propagation characteristics of the various bands.

13-5 FREQUENCY SHARING

An amateur radio operator is not assigned a specific frequency on which to operate. He operates anywhere within a band of frequencies. He shares these frequencies with other amateurs and he must make certain not to interfere with others. If he finds he is interfering with another amateur, he must work out some mutual arrangement whereby there is no interference.

In some cases, the amateurs share a band with other services. For instance, in the 1800 kHz - 2000 kHz band, the use of the frequencies is shared with the Loran–A radionavigation system. The use of these frequencies by amateur stations must not cause harmful interference to the Loran–A system. If an amateur station does cause such interference, he must cease operating on these frequencies, if so directed by the FCC.

13-6 Q SIGNAL SYSTEMS AND STANDARD ABBREVIATIONS

"Q signals" are widely used abbreviations. Their use saves time in transmitting commonly used expressions and questions. When a Q signal is followed by a question mark, it takes the form of a question. When not followed by a question mark, a Q signal is either a reply to a Q signal question, or a direct statement. For instance, "QRA?" means "What is the name of your station?" "QRA" without the question mark means "The name of my station is. . . . . ." Commonly used Q signals are listed below:

QRG? — What is my frequency?
QRM? — Are you being interfered with?
QRN? — Are you troubled by static?
QRQ — Send faster.
QRS? — Shall I send slower?
QRT — I am closing down my station.
QRU? — Do you have anything for me?
QRX — Wait or stand by.
QRZ? — Who is calling me?
QSB? — Is my signal fading?
QSL? — Can you acknowledge receipt?
QSY? — Shall I change frequency?
QSZ? — Shall I send each group twice?
QTH? — What is your location?

The following abbreviations are commonly used by amateurs in telegraphy:

CQ — A general inquiry call. It is used when an amateur wishes to make a contact.
DE — means "from". It is followed by the call letters of the station doing the sending.
K — means "go ahead". It is used after a CQ or at the end of a transmission.

AR — indicates the end of a message unit or transmission.
SK — indicates the end of the entire transmission.
R — indicates that the message has been received correctly.
73 — indicates "Best regards".

13-7 R.S.T. REPORTING SYSTEM

The RST reporting system is a means of rating the quality of a signal on a numerical basis. In this system, the R stands for readability and is rated on a scale of 1 to 5. The S stands for signal strength and it is rated on a scale of 1 to 9. The T indicates the quality of a CW tone and its scale is also 1 to 9. The higher the number, the better the signal.

READABILITY

1. Unreadable
2. Barely readable; occasional words distinguishable
3. Readable with considerable difficulty
4. Readable with practically no difficulty
5. Perfectly readable

SIGNAL STRENGTH

1. Faint; signals barely perceptible
2. Very weak signals
3. Weak signals
4. Fair signals
5. Fairly good signals
6. Good signals
7. Moderately strong signals
8. Strong signals
9. Extremely strong signals

TONE

1. Extremely rough, hissing tone
2. Very rough AC note; no trace of musicality
3. Rough, low-pitched AC note; slightly musical
4. Rather rough AC note; moderately musical
5. Musically modulated note
6. Modulated note; slight trace of whistle
7. Near DC note; smooth ripple
8. Good DC note; just a trace of ripple
9. Purest DC note (If note appears to be crystal controlled, add letter X after the number indicating tone)

EXAMPLE: Your signals are RST 599X. (Your signals are perfectly readable, extremely strong, have purest DC note, and sound as if your transmitter is crystal-controlled.)

13-8 OPERATION DURING EMERGENCIES

One of the most important responsibilities of radio amateurs is to participate, to the extent of their abilities, in providing emergency communications during times of emergency. At the very minimum, radio amateurs must yield their operating frequency if that frequency is required for emergency communications.

Radio amateurs have been instrumental in saving lives during floods, earth quakes, and accidents by using their station as an instrument of emergency communication. Hardly any period of time goes by when it is not reported in the newspapers that some radio amateur or group of amateurs have participated in the saving of lives and property during some disaster.

13-9 MESSAGE TRAFFIC

Another rewarding aspect of amateur radio is the participation of hams in handling messages on behalf of third parties. An amateur may, if he so desires, relay messages received at his station to non-amateur individuals. This is known as third party traffic. This privilege was utilized extensively during wartime when amateur operators made it possible for hundreds of thousands of servicemen to communicate with their families from overseas.

Although most message traffic is usually accomplished in an informal fashion, there are several clubs that perform this function as a public service. These clubs operate on regular schedules and handle message traffic between servicemen, students, and businessmen and their families.

Third party traffic involving any material compensation to anyone is prohibited. Third party traffic consisting of business communications is also prohibited. Third party traffic with foreign countries that do not have a third party agreement with the United States is not permitted.

13-10 NETWORK OPERATION

Network operation is a phrase applied to a group of amateurs who meet at a specified time on a specified frequency. Usually these "nets" function in one of two categories; social roundtable discussion or for the purpose of exchanging message traffic as described above. In either case, network operation represents a function of amateur radio that enables a group of amateurs to share in their common interest.

13-11 MEASURES TO PREVENT THE USE OF AMATEUR RADIO STATION EQUIPMENT BY UNAUTHORIZED PERSONS.

Some or all of the following measures can be taken to prevent the use of amateur radio station equipment by unauthorized personnel:

(1) Keep the amateur radio station room locked.

(2) Lock the power switch when the station is not in use.

(3) Post a sign in the station room or on the door to the station room indicating that only licensed authorized personnel may operate the radio station.

(4) Post the FCC licenses of those authorized to use the station in the station operating room.

PRACTICE QUESTIONS — LESSON 13

1 The abbreviation "QRM?" means:
a. Who is calling me?
b. Send faster.
c. What is my frequency?
d. Are you being interfered with?

2 The standard abbreviation for "go ahead" is:
a. CQ
b. QSY
c. K
d. AR

3 The highest priority communication is:
a. Message traffic b. Emergency c. CQ d. Network

4 Always test a transmitter:
a. on the air
b. using a dummy load
c. on a frequency outside the amateur band
d. by calling CQ

5 The abbreviation for a general call is:
a. SK b. CQ c. QRU d. Hello

6 Before operating on a certain frequency you should:
a. adjust the dummy load
b. tune up the receiver
c. check to see if the frequency is being used.
d. adjust the low pass filter

7 In communications between two points, you should use the:
a. maximum power available
b. onmidirectional antenna
c. lowest possible frequency
d. minimum power necessary

8 An RST report of 484 indicates a/an:
a. unreadable signal
b. good smooth tone
c. strong signal
d. weak signal

9 Third party traffic involves messages on behalf of a/an:
a. foreign government
b. person other than the two radio amateurs
c. another amateur
d. party in distress

10 The abbreviation for closing down a station is:
a. QRG b. QRX c. QRT d. QSZ

11 The abbreviation for sending slower is:
a. QRG. b. QRM. c. QRX. d. QRS.

12 The signal of a good operator;
a. contains audio frequencies below 6 kHz.
b. contains the maximum number of harmonics.
c. contains only even number harmonics.
d. contains a minimum amount of harmonic energy.

SECTION III – LESSON 14
RULES AND REGULATIONS

The government of the United States does not license or interfere in any way with the RECEPTION of standard broadcast or short wave programs. Anyone can own and operate a radio receiver without a license. However, in the case of radio transmission, the situation is entirely different. All transmitting stations, whether Amateur or Commercial, are licensed by the Federal Communications Commission (FCC). Consequently, all operators of transmitting equipment must be licensed.

There are various rules and regulations which govern the operation of all radio transmissions. It is important that the radio amateur be familiar with the general rules and regulations of Communications as well as those rules which apply specifically to Amateur Radio. The examinations for the Amateur Operators Licenses contain questions based on the provisions of treaties, statutes and regulations affecting amateurs. Because it is important to know the laws regarding amateur radio transmission, we are reproducing those EXCERPTS from Part 97 of the FCC rules that apply to the Novice license requirements. Part 97 contains the rules and regulations that govern the Amateur Radio Service.

Violation of the rules and regulations of the Federal Communications Commission can result in a maximum penalty of up to $500.00 for each day during which the offense occurs. The penalty can also include suspension of the operator license and revocation of the station license.

97.1 BASIS AND PURPOSE

The rules and regulations in this part are designed to provide an amateur radio service having a fundamental purpose as expressed in the following principles:

(a) Recognition and enhancement of the value of the amateur service to the public as a voluntary noncommercial communication service, particularly with respect to providing emergency communications.

(b) Continuation and extension of the amateur's proven ability to contribute to the advancement of the radio art.

(c) Encouragement and improvement of the amateur radio service through rules which provide for advancing skills in both the communication and technical phases of the art.

(d) Expansion of the existing reservoir within the amateur radio service of trained operators, technicians, and electronics experts.

(e) Continuation and extension of the amateur's unique ability to enhance international good will.

97.3 DEFINITIONS

(a) AMATEUR RADIO SERVICE. A radio communication service of self-training, intercommunication, and technical investigation carried on by amateur radio operators.

(b) AMATEUR RADIO COMMUNICATION. Noncommercial radio communication by or among amateur radio stations solely with a personal aim and without pecuniary or business interest.

(c) AMATEUR RADIO OPERATOR means a person holding a valid license to operate an amateur radio station issued by the Federal Communications Commission.

(d) AMATEUR RADIO LICENSE. The instrument of authorization issued by the Federal Communications Commission comprised of a station license, and in the case of the primary station, also incorporating an operator license.

OPERATOR LICENSE. The instrument of operator authorization including the class of operator privileges.

STATION LICENSE. The instrument of authorization for a radio station in the amateur radio service.

(e) AMATEUR RADIO STATION. A station licensed in the amateur radio service embracing necessary apparatus at a particular location used for amateur radio communication.

(f) PRIMARY STATION. The principal amateur radio station at a specific land location shown on the station license.

(p) CONTROL OPERATOR. An amateur radio operator designated by the licensee of an amateur radio station to also be responsible for the emissions from that station.

(w) THIRD-PARTY TRAFFIC. Amateur radio communication by or under the supervision of the control operator at an amateur radio station to another amateur radio station on behalf of anyone other than the control operator.

97.5 CLASSES OF OPERATOR LICENSES.

Amateur Extra Class.
Advanced Class (previously Class A).
General Class (previously Class B).
Conditional Class (previously Class C).
Technician Class.
Novice Class.

97.7 PRIVILEGES OF AMATEUR OPERATOR LICENSES. (See chart on Page 124 showing privileges of operator licenses for popular amateur bands.)

(a) AMATEUR EXTRA CLASS AND ADVANCED CLASS. All authorized amateur privileges including exclusive frequency operating authority in accordance with the following table.

Frequencies	Class of license authorized.
3500 - 3525 kHz 3775 - 3800 kHz 7000 - 7025 kHz 14, 000 - 14, 025 kHz 21, 000 - 21, 025 kHz 21, 250 - 21, 270 kHz	Amateur Extra Only.
3800 - 3890 kHz 7150 - 7225 kHz 14, 200 - 14, 275 kHz 21, 270 - 21, 350 kHz	Amateur Extra and Advanced.

(b) GENERAL CLASS. All authorized amateur privileges except those exclusive operating privileges which are reserved to the Advanced Class and/or Amateur Extra Class.

(c) CONDITIONAL CLASS. Same privileges as General Class. New Conditional Class licenses will not be issued. Present Conditional Class licensees will be issued General Class licenses at time of renewal or modification.

(d) TECHNICIAN CLASS. All authorized amateur privileges on the frequencies 50.0 MHz and above. Technician Class licenses also convey the full privileges of Novice Class licenses.

(e) NOVICE CLASS. Radiotelegraphy in the frequency bands 3700 - 3750 kHz, 7100 - 7150 kHz (7050 - 7075 kHz when the terrestrial station location is not within Region 2), 21,100 - 21,200 kHz, and 28,100 - 28,200 kHz, using only Type A1 emission.

97.9 ELIGIBILITY FOR NEW OPERATOR LICENSE.

Anyone, except a representative of a foreign government, is eligible for an amateur operator license.

97.13 RENEWAL OR MODIFICATION OF AMATEUR OPERATOR LICENSE

(a) An amateur radio operator license may be renewed upon proper application.

(b) The applicant shall qualify for a new license by examination if the requirements of this section are not fulfilled.

97.28 MANNER OF CONDUCTING EXAMINATIONS

(a) Except as provided by 97.27, all examinations for Amateur Extra, Advanced, General and Technician Class operator licenses will be conducted by authorized Commission personnel or representatives at locations and times specified by the Commission. Examination elements given under the provisions of 97.27 will be administered by an examiner selected by the Commission. All applications for consideration of eligibility under 97.27 should be filed on FCC Form 610, and should be sent to the FCC field office nearest the applicant. (A list of these offices appears in Sec. 0.121 of the Commission's Rules and can be obtained from the Regional Services Division, Field Operations Bureau, FCC, Washington, D.C. 20554, or any field office.)

(b) Unless otherwise prescribed by the Commission, examinations for the Novice Class license will be conducted and supervised by a volunteer examiner selected by the applicant. The volunteer examiner shall be at least 18 years of age, shall be unrelated to the applicant, and shall be the holder of an Amateur Extra, Advanced or General Class operator license. The written portion of the Novice examination, Element 2, shall be obtained administered, and submitted in accordance with the following procedure:

(1) Within 10 days after successfully completing telegraphy examination element 1(a), an applicant shall submit an application (FCC Form 610) to the Commission's office in Gettysburg, Pennsylvania 17325. The application shall include a written request from the volunteer examiner for the examination papers for Element 2. The examiner's written request shall include (i) the names and permanent addresses of the examiner and the applicant, (ii) a description of the examiner's qualifications to administer the examination, (iii) the examiner's statement that the applicant has passed telegraphy element 1(a) under his supervision within the 10 days prior to submission of the request, and (iv) the examiner's written signature. Exam-

ination papers will be forwarded only to the volunteer examiner.

(2) The volunteer examiner shall be responsible for the proper conduct and necessary supervision of the examination. Administration of the examination shall be in accordance with the instructions included with the examination papers.

97.37 ELIGIBILITY FOR STATION LICENSE.

(a) Except as provided in paragraphs (b) and (c) of this section, an amateur radio station license shall be issued only to a licensed amateur radio operator.

(b) A military recreation station license may be issued to, (1) a licensed amateur radio operator; or (2) an individual who is in charge of a proposed military recreation station and who is not a representative of a foreign government.

(c) A club station license may be issued to an amateur radio club if, (1) the club meets the definition of CLUB STATION in P.97.3; (2) the club demonstrates to the FCC a compelling need for a club station license; and (3) at least one officer of the club holds an amateur operator license of the Technician Class or above.

(d) An amateur radio station license shall not be issued to a school, company, corporation, association or other organization, except to an amateur radio club meeting the requirements in paragraph (c) of this section and in P.97.3.

97.39 STATION LICENSE REQUIRED

(a) No transmitting station shall be operated in the amateur radio service without being licensed by the Federal Communications Commission.

(b) Every amateur radio operator shall have one, but only one, primary amateur radio station license.

97.45 LIMITATIONS ON ANTENNA STRUCTURES

(a) Except as provided in paragraph (b) of this section, an antenna for a station in the Amateur Radio Service which exceeds the following height limitations, may not be erected or used unless notice has been filed with both the FAA on FAA Form 7460-1 and with the Commission on Form 714 or on the license application form, and prior approval by the Commission has been obtained for:

(1) Any construction or alteration of more than 200 feet in height above ground level at its site (sec. 17.7(a) of this chapter).

(2) Any construction or alteration of greater height than an imaginary surface extending outward and upward at one of the following slopes (Sec. 17.7(b) of this chapter);

(i) 100 to 1 for a horizontal distance of 20,000 feet from the nearest point of the nearest runway of each airport, with at least one runway more than 3,200 feet in length, excluding heliports and seaplane bases without specified boundaries, if that airport is either listed in the Airport Directory of the current Airman's Information Manual or is operated by a Federal military agency.

(ii) 50 to 1 for a horizontal distance of 10,000 feet from the nearest point of the nearest runway of each airport with its longest runway no more than 3,200 feet in length, excluding heliports and seaplane bases without specified boundaries, if that airport is either listed in the Airport Directory or is operated by a Federal military agency.

(iii) 25 to 1 for a horizontal distance of 5,000 feet from the nearest point of the nearest landing and takeoff area of each heliport listed in the Airport Directory or operated by a Federal military agency.

(3) Any construction or alteration on an airport listed in the Airport Directory of the Airman's Information Manual (Sec. 17.7(c) of this chapter).

(b) A notification to the Federal Aviation Administration is not required for any of the following construction or alteration:

(1) Any object that would be shielded by existing structures of a permanent and substantial character or by natural terrain or topographic features of equal or greater height, and would be located in the congested area of a city, town, or settlement where it is evident beyond all reasonable doubt, that the structure so shielded will not adversely affect safety in air navigation. Applicants claiming such exemption shall submit a statement with their application to the Commission explaining the basis in detail for their finding (Sec. 17.14(a) of this chapter).

(2) Any antenna structure of 20 feet or less in height, except one that would increase the height of another antenna structure (Sec. 17.14(b) of this chapter).

(c) Further details as to whether an aeronautical study and/or obstruction marking and lighting may be required, and specifications for obstruction marking and lighting when required, may be obtained from Part 17 of this chapter, "Construction, Marking and Lighting of Antenna Structures". Information regarding the inspection and maintenance of antenna structures requiring obstruction marking and lighting, is also contained in Part 17 of this chapter.

97.49 COMMISSION MODIFICATION OF STATION LICENSE.

(a) Whenever the Commission shall determine that public interest, convenience and necessity would be served, or any treaty ratified by the United States will be more fully complied with, by the modification of any radio station license either for a limited time, or for the duration of the term thereof, it shall issue an order for such licensee to show cause why such license should not be modified.

(b) Such order to show cause shall contain a statement of the grounds and reasons for such proposed modification, and shall specify wherein the said license is required to be modified. It shall require the licensee against whom it is directed to appear at a place and time therein named, in no event to be less than 30 days from the date of receipt of the order, to show cause why the proposed modification should not be made and the order of modification issued.

(c) If the licensee against whom the order to show cause is directed does not appear at the time and place provided in said order, a final order of modification shall issue forthwith.

97.59 LICENSE TERM

Amateur operator licenses and station licenses are normally valid for a period of 5 years from the date of issuance of a new or renewed license. All station licenses, regardless of when issued, will expire on the same date as the licensee's operator license.

97.67 MAXIMUM AUTHORIZED POWER

(b) Notwithstanding other provisions, amateur stations shall use the minimum amount of transmitter power necessary to carry out the desired communications.

(d) In the frequency bands 3700 - 3750 kHz, 7100 - 7150 kHz (7050 - 7075 kHz when the terrestrial location of the station is not within Region 2) 21, 100 - 21, 200 kHz, and 28, 100 - 28, 200 kHz, the power input to the transmitter final amplifying stage supplying radio frequency energy to the antenna shall not exceed 250 watts, exclusive of power for heating the cathode of a vacuum tube(s).

97.74 FREQUENCY MEASUREMENT AND REGULAR CHECK

The licensee of an amateur station shall provide for measurement of the emitted carrier frequency or frequencies and shall establish procedure for making such measurements regularly. The measurement of the emitted carrier frequency or frequencies shall be made by means independent of the means used to control the radio frequency or frequencies generated by the transmitting apparatus and shall be of sufficient accuracy to assure operating within the amateur frequency band used.

97.78 PRACTICE TO BE OBSERVED BY ALL LICENSEES

In all aspects not specifically covered by these regulations, each amateur station shall be operated in accordance with good engineering and good amateur practice.

97.79 CONTROL OPERATOR REQUIREMENTS

(a) The licensee of an amateur station shall be responsible for its proper operation.

(b) Every amateur radio station, when in operation, shall have a control operator at an authorized control point. The control operator shall be on duty, except where the station is operated under automatic control. The control operator may be the station licensee, if a licensed amateur radio operator, or may be another amateur radio operator with the required class of license and designated by the station licensee. The control operator shall also be responsible, together with the station licensee, for the proper operation of the station.

(c) An amateur station may only be operated in the manner and to the extent permitted by the operator privileges authorized for the class of license held by the control operator, but may exceed those of the station licensee provided proper station identification procedures are performed.

(d) The licensee of an amateur radio station may permit any third party to participate in amateur radio communication from his station, provided that a control operator is present and continuously monitors and supervises the radio communication to insure compliance with the rules.

97.82 AVAILABILITY OF OPERATOR LICENSE

The original license of each operator shall be kept in the personal possession of the operator while operating an amateur station. When operating an amateur station at a fixed location, however, the license may be posted in a conscipuous place in the room occupied by the operator. The license shall be available for inspection by an authorized government official whenever the operator is operating an amateur station and, at other times, upon request made by an authorized representative of the Commission, except when such license has been filed with application for modification or renewal thereof, or has been mutilated, lost or destroyed, and application has been made for a duplicate license in accordance with 97.57. Photo copies of the license can be made but cannot be used in lieu of the original as required by this section.

97.83 AVAILABILITY OF STATION LICENSE

The original license of each amateur station or a photocopy thereof shall be posted in a conspicuous place in the room occupied by the licensed operator while the station is being operated at a fixed location, or shall be kept in his personal possession. When the station is operated at other than a fixed location, the original station license or a photocopy thereof shall be kept in the personal possession of the station licensee (or a licensed representative) who shall be present at the station while it is being operated as a portable or mobile station. The original station license shall be available for inspection by any authorized Government official at all times while that station is being operated, and at other times upon request made by an authorized representative of the Commission, except when such license has been filed with application for modification or renewal thereof, or has been mutilated, lost, or destroyed, and request has been made for a duplicate license in accordance with §97.57.

97.84 STATION IDENTIFICATION

(a) An amateur station shall be identified by the transmission of its call sign at the beginning and end of each single transmission or exchange of transmissions and at intervals not to exceed 10 minutes during any single transmission or exchange of transmissions of more than 10 minutes duration. Additionally, at the end of an exchange of telegraphy (other than teleprinter) or telephony transmissions between amateur stations, the call sign (or the generally accepted network identifier) shall be given for the station, or for at least one of the group of stations, with which communication was established.

(b) Under conditions when the control operator is other than the station licensee, the station identification shall be the assigned call sign for that station. However, when a station is operated within the privileges of the operator's class of license but which exceeds those of the station licensee, station identification shall be made by following the station call sign with the operator's primary station call sign (i.e. WN4XYZ/W4XX).

97.89 POINTS OF COMMUNICATIONS

(a) Amateur stations may communicate with:

(1) Other amateur stations, excepting those prohibited by Appendix 2. (2) Stations in other services licensed by the FCC and with US Government stations for civil defense purposes in accordance with Subpart F, in emergencies, and on a temporary basis, for test purposes. (3) Any station which is authorized by the FCC to communicate with amateur stations.

(b) Amateur stations may be used for transmitting signals, or communications, or energy, to receiving apparatus for the measurement of emissions, temporary observation of transmission phenomena, radio control of remote objects, and similar experimental purposes and for the purposes set forth in §97.91.

97.91 ONE-WAY COMMUNICATIONS

In addition to the experimental one-way transmission permitted by Section 97.89, the following kinds of one-way communications addressed to amateur stations, are authorized and will not be construed as broadcasting:

(a) Emergency communications, including bonafide emergency drill practice transmissions;

(b) Information bulletins consisting solely of subject matter having direct interest to the amateur radio service as such;

(c) Round-table discussions or net-type operations where more than two amateur stations are in communication, each station taking a turn at transmitting to other station(s) of the group; and

(d) Code practice transmissions intended for persons learning or improving proficiency in the International Morse Code.

97.103 STATION LOG REQUIREMENTS

An accurate legible account of station operation shall be entered into a log for each amateur radio station. The following items shall be entered as a minimum:

(a) The call sign of the station, the signature of the station licensee, or a photocopy of the station license.

(b) The locations and dates upon which fixed operation of the station was initiated and terminated. If applicable, the location and dates upon which portable operation was initiated and terminated at each location.

(1) The date and time periods the duty control operator for the station was other than the station licensee, and the signature and primary station call sign of that duty control operator.

(2) A notation of third party traffic sent or received, including names of all third parties, and a brief description of the traffic content. This entry may be in a form other than written, but one which can be readily transcribed by the licensee into written form.

(3) Upon direction of the Commission, additional information as directed shall be recorded in the station log.

97.105 RETENTION OF LOGS

The station log shall be preserved for a period of at least 1 year following the last date of entry and retained in the possession of the licensee. Copies of the log, including the sections required to be transcribed by 97.103, shall be available to the Commission for inspection.

97.112 NO REMUNERATION FOR USE OF STATION

(a) An amateur station shall not be used to transmit or receive messages for hire, nor for communication for material compensation, direct or indirect, paid or promised.

97.113 BROADCASTING PROHIBITED

Subject to the provisions of 97.91 of these rules, an amateur station shall not be used to engage in any form of broadcasting, that is, the dissemination of radio communications intended to be received by the public directly or by the intermediary of relay stations, nor for the retransmission by automatic means of programs or signals emanating from any class of station other than amateur. The foregoing provision shall not be construed to prohibit amateur operators from giving their consent to the rebroadcast by broadcast stations of the transmissions of their amateur stations, provided, that the transmissions of the amateur stations shall not contain any direct or indirect reference to the re-broadcast.

97.114 THIRD PARTY TRAFFIC

The transmission or delivery of the following amateur radio-communication is prohibited:

(a) International third party traffic except with countries which have assented thereto;

(b) Third party traffic involving material compensation, either tangible or intangible, direct or indirect, to a third party, a station licensee, a control operator, or any other person.

(c) Except for an emergency communication as defined in this part, third party traffic consisting of business communications on behalf of any party. For the purpose of this section business communication shall mean any transmission or communication the purpose of which is to facilitate the regular business or commercial affairs of any party.

97.119 OBSCENITY, INDECENCY, PROFANITY

No licensed radio operator or other person shall transmit communications containing obscene, indecent or profane words, language or meaning.

97.121 FALSE SIGNALS

No licensed radio operator shall transmit false or deceptive signals or communications by radio, or any call letter or signal which has not been assigned by proper authority to the radio station he is operating.

97.123 UNIDENTIFIED COMMUNICATIONS

No licensed radio operator shall transmit unidentified radio communications or signals.

97.125 INTERFERENCE

No licensed radio operator shall willfully or maliciously interfere with or cause interference to any radio communication or signal. (The penalty for violation of this rule is as follows: A fine of up to $500.00 for each day during which the offense occurs, and suspension of the operator's license. In case the interference is in connection with distress communications, the penalty may be a maximum fine of $10,000.00 or imprisonment up to one year, or both, and the revocation of the station license).

97.131 RESTRICTED OPERATION

(a) If the operation of an amateur station causes general interference to the reception of transmissions from stations operating in the domestic broadcast service when receivers of good engineering design including adequate selectivity characteristics are used to receive such transmissions, and this fact is made known to the amateur station licensee, the amateur station shall not be operated during the hours from 8 p.m. to 10:30 p.m., local time, and on Sunday for the additional period from 10:30 a.m. until 1 p.m., local time, upon the frequency or frequencies used when the interference is created.

(b) In general, such steps as may be necessary to minimize interference to stations operating in other services may be required after investigation by the Commission.

97.133 SECOND NOTICE OF SAME VIOLATION

In every case where an amateur station licensee is cited within a period of 12 consecutive months for the second violation of the provisions of §§97.61, 97.63, 97.65, 97.71, or 97.73, the station licensee, if directed to do so by the Commission, shall not operate the station and shall not permit it to be operated from 6 p.m. to 10:30 p.m., local time, until written notice has been received authorizing the resumption of full-time operation. This notice will not be issued until the licensee has reported on the results of tests which he has conducted with at least two other amateur stations at hours other than 6 p.m. to 10:30 p.m., local time. Such tests are to be made for the specific purpose of aiding the licensee in determining whether the emissions of the station are in accordance with the Commission's rules. The licensee shall report to the Commission the observations made by the cooperating amateur licensees in relation to the reported violations. This report shall include a statement as to the corrective measures taken to insure compliance with the rules.

97.135 THIRD NOTICE OF SAME VIOLATION

In every case where an amateur station licensee is cited within a period of 12 consecutive months for the third violation of § §97.61, 97.63, 97.65, 97.71, or 97.73, the station licensee, if directed by the Commission, shall not operate the station and shall not permit it to be operated from 8 a.m. to 12 midnight, local time, except for the purposes of transmitting a prearranged test to be observed by a monitoring station of the Commission to be designated in each particular case. The station shall not be permitted to resume operation during these hours until the licensee is authorized by the Commission, following the test, to resume full-time operation. The results of the test and the licensee's record shall be considered in determining the advisability of suspending the operator license or revoking the station license, or both.

97.137 ANSWERS TO NOTICES OF VIOLATIONS

Any licensee receiving official notice of a violation of the terms of the Communications Act of 1934, as amended, any legislative act, Executive order, treaty to which the United States is a party, or the rules and regulations of the Federal Communications Commission, shall, within 10 days from such receipt, send a written answer direct to the office of the Commission originating the official notice: Provided, however, That if an answer cannot be sent or an acknowledgment made within such 10-day period by reason of illness or other unavoidable circumstances, acknowledgment and answer shall be made at the earliest practicable date with a satisfactory explanation of the delay. The answer to each notice shall be complete in itself and shall not be abbreviated by reference to other communications or answers to other notices. If the notice relates to some violation that may be due to the physical or electrical characteristics of transmitting apparatus, the answer shall state fully what steps, if any, are taken to prevent future violations, and if any new apparatus is to be installed, the date such apparatus was ordered, the name of the manufacturer, and promised date of delivery. If the notice of violation relates to some lack of attention to or improper operation of the transmitter, the name of the operator in charge shall be given.

CALL SIGNS

The amateur call sign consists of three parts: (1) the prefix, (2) a single digit, and (3), a suffix.

The prefix can be one or two letters. Single letters are either K, N, or W. Two letter combinations are either AA through AL, KA through KZ, NA through NZ, or WA through WZ.

The digit is a single number, Ø through 9, indicating a geographical district.

The suffix can be one, two or three letters. Single letters are A through Z. Two letter combinations are AA through ZZ. Three letter combinations are AAA through ZZZ.

PRACTICE QUESTIONS — LESSON 14

1 One of the fundamental purposes of amateur radio is not:
  a. enhancing international good will.
  b. expanding the reservoir of electronics experts.

c. advancing technical skills of radio.
d. encouraging others to become radio amateurs.

2 In the event of a violation of the Rules at a transmitter operated by a control operator other than the station licensee, who is responsible?
a. the licensee.
b. the control operator.
c. the licensee and the control operator.
d. the owner of the station.

3 Which one of the following types of one-way communication is not prohibited?
a. music
b. broadcasting
c. round table discussions
d. unidentified communications

4 For how long must a log be preserved?
a. 6 months
b. 12 months
c. 2 years
d. 3 years

5 The control operator:
a. is designated by the FCC.
b. may be designated by the station licensee.
c. must have a higher class of license than the station licensee
d. is not responsible for the emissions from the station.

6 A log need not contain:
a. location of operation.
b. notation of third party traffic.
c. the operating frequency.
d. dates of operation.

7 Radio messages having top priority are:
a. relief or emergency messages
b. messages sent to foreign amateurs
c. ordinary calls
d. ship to shore messages

8 The maximum input power permitted to the final stage of a transmitter, owned and operated by a Novice operator, is:
a. 50 watts
b. 75 watts
c. 100 watts
d. 250 watts

9 Which of the following is true for a Novice license?
a. it is valid for 5 years
b. it is valid for 1 year
c. it may not be renewed
d. it is valid for 3 years

10 Station identification must be transmitted at intervals not to exceed:
a. 30 minutes
b. 15 minutes
c. 10 minutes
d. one hour

11 One way communications are not allowed for;
a. emergency communications
b. authorized code practice transmissions
c. broadcasting music
d. round-table discussions

12 Third party traffic is allowed for:
a. emergency communication
b. profit
c. business.
d. profit for the amateur.

13 The penalty for willful interference with other radio communications is:
a. restricting operation to the 20 meter band
b. restricting operator to telegraph operation
c. fine and suspension of license
d. restriction to local calls

14 A state of emergency affecting amateur communications becomes effective when:
a. an emergency occurs
b. so ordered by the FCC
c. at the discretion of the operator
d. 3 hours after the emergency has started

FINAL NOVICE CLASS "FCC-TYPE" EXAMINATION

1. What part of the FCC rules govern the Amateur Radio Service?
A. 95 B. 97 C. 73 D. 74

2. A control operator of a station:
A. must be designated by the FCC
B. must be designated by the engineer-in-charge of the local FCC radio district office
C. may be designated by the station licensee
D. must hold a General Class or higher grade of license

3. Which of the following is not prohibited?
A. unidentified communications C. profanity
B. broadcasting D. round table discussions

4. Which of the following frequencies cannot be used by a Novice class operator?
A. 3741 kHz B. 7152 kHz C. 21,101 kHz D. 28,101 kHz

5. Which of the following indicates the end of the entire transmission?
A. K B. SK C. DE D. AR

6. Which of the following reports indicates a moderately strong signal that is readable, but with considerable difficulty?
A. R2S8 B. R2S7 C. R4S8 D. R3S7

7. What is the unit of capacitance?
A. henry B. farad C. ohm D. volt

8. The abbreviaton for kiloHertz is:
A. kT B. kH C. kZ D. kHz

9. AC is changed to DC by means of a:
A. rectifier B. transistor C. filter D. transformer

10. What is the total resistance of a circuit containing a 12 ohm resistor in parallel with a 24 ohm resistor?
A. 12 ohms B. 36 ohms C. 2 ohms D. 8 ohms

11. What is the meaning of QRS?
A. what is your signal strength?
B. am I being interfered with?
C. what is your location?
D. shall I transmit more slowly?

12. What is the value of a resistor having a current of 3 amperes flowing through it and 30 volts across it?
A. 10 ohms B. .1 ohm C. 90 ohms D. 270 ohms

13. An advantage of a transmatch is NOT:
A. a reduction of harmonics
B. matches transmitter to antenna system
C. amplifies RF from final amplifier
D. feeds maximum energy to antenna

14. An advantage of a multiband antenna is:
A. low SWR
B. high efficiency
C. high directivity
D. several bands can be operated with one antenna

15. A Novice Class operator may operate an amateur radio station in the following band:
A. 3700 kHz - 3750 kHz
B. 50.0 MHz - 54.0 MHz
C. 21.0 MHz - 21.25 MHz
D. 145.0 MHz - 147.0 MHz

16. One kiloHertz is equal to:
A. 10,000 Hertz
B. 1000 Hertz
C. 1000 megaHertz
D. 100,000 Hertz

17. The second harmonic of 400 kHz. is:
A. 400 c
B. 800 c
C. 800 kHz
D. 1200 kHz

18. Interference due to sparking at the telegraph key contacts can be eliminated by a:
A. spark suppressor
B. resistor spark plug
C. key-click filter
D. spark gap device

19. An oscillator does not contain a/an;
A. tuned circuit
B. relay
C. tetrode
D. method of feedback

20. The ideal SWR is:
A. 10:1
B. 5:1
C. 2:1
D. 1:1

21. Which of the following elements are exclusive to a transistor?
A. grid, plate, emitter
B. base, cathode, collector
C. base, emitter, collector
D. base, cathode, collector

22. A CW transmitter does NOT contain a/an:
A. detector
B. oscillator
C. RF Amplifier
D. power amplifier

23. The weak output signal of a microphone is made larger by means of a/an:
A. detector
B. AF amplifier
C. RF amplifier
D. speaker

24. Power is measured by a/an:
A. wattmeter
B. frequency meter
C. watt-hour meter
D. voltmeter

ADDITIONAL INFORMATION

Rules and Regulations

The group D format call sign consists of a two letter prefix (starting with K or W), a single digit (0 through 9), and a three letter suffix (AAA through ZZZ).

The call sign prefixes for Amateur Radio stations licensed by the FCC can be one or two letters. Single letters are either K, N, or W. Two letter combinations are either AA through AL, KA through KZ, NA through NZ, or WA through WZ.

The term used in the FCC rules to describe signals being transmitted to receiving apparatus while in beacon or radio control operation is "one way transmission".

Amateur Radio Practice

Some or all of the following measures should be taken to prevent the use of amateur radio station equipment by unauthorized persons:

(1) Keep the amateur radio station room locked.
(2) Lock the power switch when the station is not in use.
(3) Install a hidden power switch that must be turned on in order to operate the station.
(4) Post a sign in the station room, or on the door of the station room, indicating that only licensed authorized personnel may operate the radio station.
(5) Post the FCC licenses of those authorized to use the station in the operating room.

When working on an antenna, mounted on a tower, a person doing the climbing should wear a good quality safety belt.

Practical Circuits

Figure 2G-1.1 illustrates a block diagram of a crystal controlled transmitter.

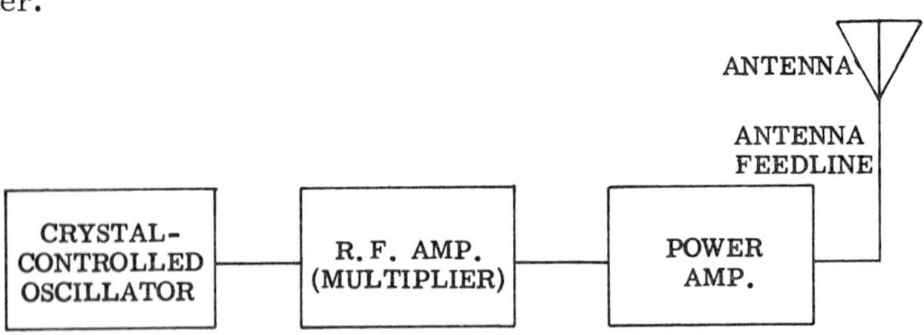

Fig. 2G-1.1. A crystal controlled transmitter.

The crystal controlled oscillator generates a single fixed frequency. If we wish to change the frequency, we must change the crystal to one of a different frequency. The RF output of the oscillator is then fed to an RF amplifier, which amplifies the RF output of the oscillator. The RF amplifier may also serve as a frequency multiplier. The frequency multiplier multiplies the frequency of the signal up to the desired frequency. The RF amplifier also acts as a buffer between the power amplifier and the oscillator. It prevents changes in the power amplifier from affecting the oscillator. This improves the stability of the transmitter. The power amplifier takes the signal from the RF amplifier and amplifies it in terms of power. The signal is then fed to the antenna where it is radiated.

A transmitter need not have three stages. The oscillator alone can act as a transmitter. However, the more stages the transmitter has, the better it is.

Figure 2G-1.3 illustrates a transmitter with a variable frequency oscillator.

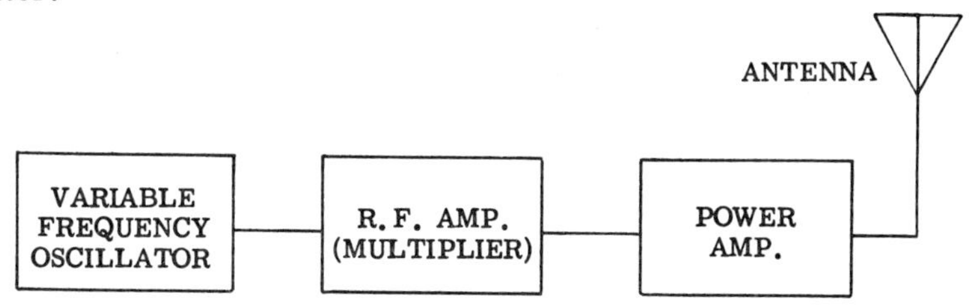

Fig. 2G-1.3 A variable frequency transmitter.

This transmitter is similar in all respects to the transmitter of Figure 2G-1.1, except that a variable frequency oscillator has replaced the crystal controlled oscillator. The variable frequency oscillator allows the operator to change his frequency by means of a knob instead of having to change crystals. This allows for easier operating. While the variable frequency oscillator is not as stable as a crystal controlled oscillator, it is quite satisfactory.

Signals and Emissions

A common cause of superimposed hum is a defective filter capacitor in the power supply. Replacing the defective capacitor will remove the hum from the signal.

APPENDIX I

STUDY GUIDE FOR ELEMENT 2 EXAMINATION FOR NOVICE CLASS AMATEUR RADIO OPERATOR LICENSE

A. RULES AND REGULATIONS

A. 2.1. BASIS AND PURPOSE: (1) Voluntary non-commercial communications service, (2) Advancement of the radio art, (3) Creation of a reservoir of trained radio operators and electronic experts.

A. 2.2. DEFINITIONS: (1) Amateur Radio Service, (2) Amateur radio operator, (3) Amateur radio station, (4) Control operator, (5) Station license, (6) Primary station.

A. 2.3. NOVICE CLASS OPERATOR PRIVILEGES: (1) Frequencies, (2) Emissions, (3) Transmitter power.

A. 2.4. LIMITATIONS: (1) License period, (2) Antenna structures.

A. 2.5. RESPONSIBILITIES: (1) Station licensee, (2) Third Party, (3) Control operator.

A. 2.6. STATION OPERATION: (1) Station identification, (2) One-way communications, (3) Operator license availability, (4) Station license availability, (5) Station logs, (6) Frequency measurement, (7) Points of communication.

A. 2.7. ADMINISTRATIVE SANCTIONS: (1) Notice of violation (2) Restricted operation.

A. 2.8. PROHIBITED PRACTICES: (1) Broadcasting, (2) Unidentified communications, (3) Obscenity, indecency and profanity, (4) Interference, (5) Third party traffic.

A. 2.9. LICENSES: (1) General eligibility: (a) Operator, (b) Station; (2) Renewal, (3) Commission Modification, (4) Availability and posting.

B. RADIO PHENOMENA

B. 2.1. DEFINITIONS: (1) Sky wave, (2) Ground wave, (3) Refraction, (4) Sunspot cycle, (5) Skip distance, (6) Wavelength, (7) Ionosphere.

B. 2.2. WAVE PROPAGATION: (1) Types of propagation: (a) Skywave versus ground wave; (2) Atmospheric conditions versus communications: (a) Daylight versus night hours, (b) Seasonal variations, (c) Ionospheric storms; (3) Effects of ionization upon wave propagation, (4) Wavelength versus frequency, (5) Frequency versus distance, (6) Velocity of radio waves.

C. OPERATING PROCEDURES

C. 2.1. BASIC PRINCIPLES: (1) Courtesy, (2) Frequency selection, (3) Frequency sharing, (4) Avoiding interference.

C. 2.2. TELEGRAPHY PROCEDURES: (1) Q signal system, (2) R.S.T. reporting system, (3) Standard abbreviations, (4) Choice of code speed.

C. 2.3. PUBLIC SERVICE OPERATING: (1) Responsibility, (2) Message traffic, (3) Network operation, (4) Operation during emergencies.

D. EMISSION CHARACTERISTICS

D. 2.1. DEFINITIONS: (1) Spurious emissions, (2) Key Clicks, (3) Chirps, (4) Carrier frequency, (5) Frequency drift, (6) Continuous waves.

D. 2.2. CLASSIFICATION OF EMISSIONS: (1) AØ, (2) A1.

D. 2.3. GENERAL FACTORS CONCERNING A1 EMISSIONS: (1) Standards of good quality A1 emissions, (2) Methods of keying, (3) Frequency stability, (4) Monitoring the transmitted signal.

E. ELECTRICAL PRINCIPLES

E. 2.1. DEFINITIONS: (1) Electromotive force, (2) Resistance, (3) Capacitance, (4) Inductance, (5) Alternating current, (6) Hertz, kilohertz & megahertz, (7) Direct current, (8) Voltage drop, (9) Electrical power, (10) Rectification, (11) Spurious signals, (12) Harmonics.

E. 2.2. FUNDAMENTAL UNITS AND CONCEPTS: (1) Units: (a) Volt, (b) Ampere, (c) Ohm, (d) Watt, (e) Henry, (f) Farad; (2) Potential difference, (3) Electrical energy.

E. 2.3. DIRECT CURRENT THEORY: (1) Ohm's Law: (a) Fundamental calculations: (i) Current, (ii) Voltage, (iii) Power; (b) Resistance: (i) In series, (ii) In parallel; (2) Series and parallel circuits: (i) Current through the branches, (ii) Voltage across the branches.

E. 2.4. PRINCIPLES OF MAGNETISM: (1) Fundamental laws: (a) Magnetic fields, (b) Magnetomotive force.

F. PRACTICAL CIRCUITS

F. 2.1. BASIC CIRCUITS: (1) Elementary oscillators, (2) Elementary amplifiers, (3) Elementary transmitters, (4) Elementary receivers.

F. 2.2. FILTER CIRCUITS: (1) Types versus utilization: (a) Key click, (b) Low pass, (c) High pass.

F. 2.3. SOLID STATE AND VACUUM TUBE RECTIFIER CIRCUITS: (1) Types versus utilization: (a) Half wave, (b) Full wave.

G. CIRCUIT COMPONENTS

G. 2.1. COMPONENT PARTS: (1) Types characteristics, applications, and schematic symbols: (a) Capacitors, (b) Crystals, (c) Transformers, (d) Resistors, (e) Diodes, (f) Transistors, (g) Inductors.

G. 2.2. VACUUM TUBES: (1) Classification by elements, (2) Basic operating principles.

G. 2.3. METERS: (1) Voltmeter, (2) Ammeter, (3) Ohmmeter.

H. ANTENNAS AND TRANSMISSION LINES

H. 2.1. DEFINITIONS: (1) Electrical length, (2) Antenna input impedance, (3) Characteristic impedance, (4) Standing waves, (5) Transmatch.

H. 2.2. DIPOLE ANTENNA: (1) Basic characteristics, (2) Length versus frequency, (3) Harmonic operation.

H. 2.3. TRANSMISSION LINES: (1) Types: (a) Open-wire parallel conductor, (b) Single wire line, (c) Coaxial; (2) Standing wave ratio: (a) Significance, (b) Measurement.

I. RADIO COMMUNICATION PRACTICES

I. 2.1. DEFINITIONS: (1) Ground Rod, (2) Ground potential, (3) Amplifier plate circuit.

I. 2.2. RADIO FREQUENCY INTERFERENCE: (1) Television receiver interference, (2) Audio equipment interference, (3) Harmonic radiation, (4) Interference to other devices.

I. 2.3. USE OF TEST EQUIPMENT: (1) Voltmeter, (2) Ammeter, (3) Ohmmeter, (4) Wattmeter.

I. 2.4. TRANSMITTERS: (1) Determination of transmitter power, (2) Determination of transmitter frequency, (3) Tuning and loading, (4) Harmonic tests and reduction.

I. 2.5. SAFETY: (1) Electrical shock avoidance, (2) Lightning protection, (3) Treatment for electrical shock, (4) Antenna support installation and maintenance.

APPENDIX 2
COMMON RADIO SYMBOLS

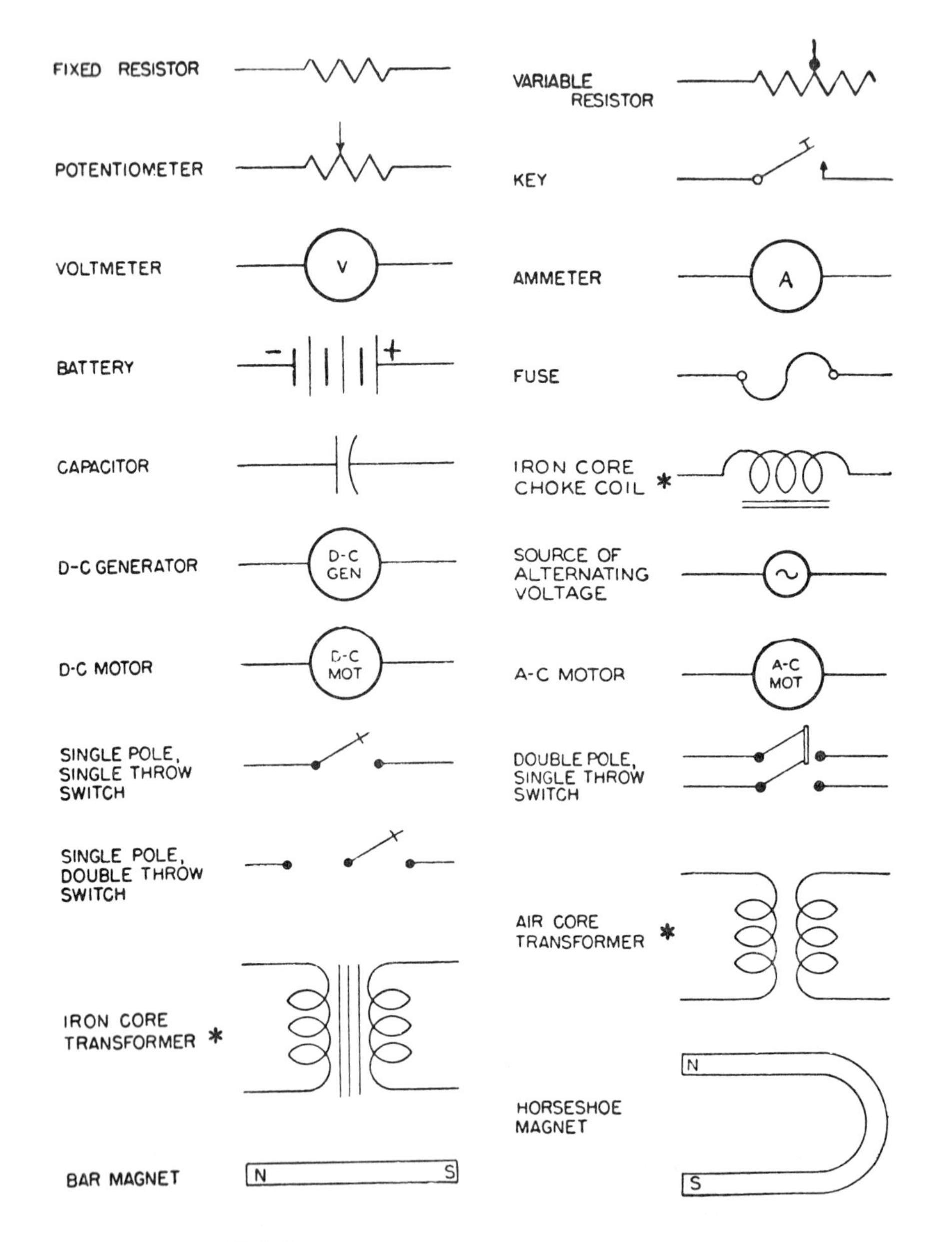

*Either (coil symbol) or (coil symbol) may be used to represent coils of wire on chokes or transformers.

APPENDIX 2
COMMON RADIO SYMBOLS

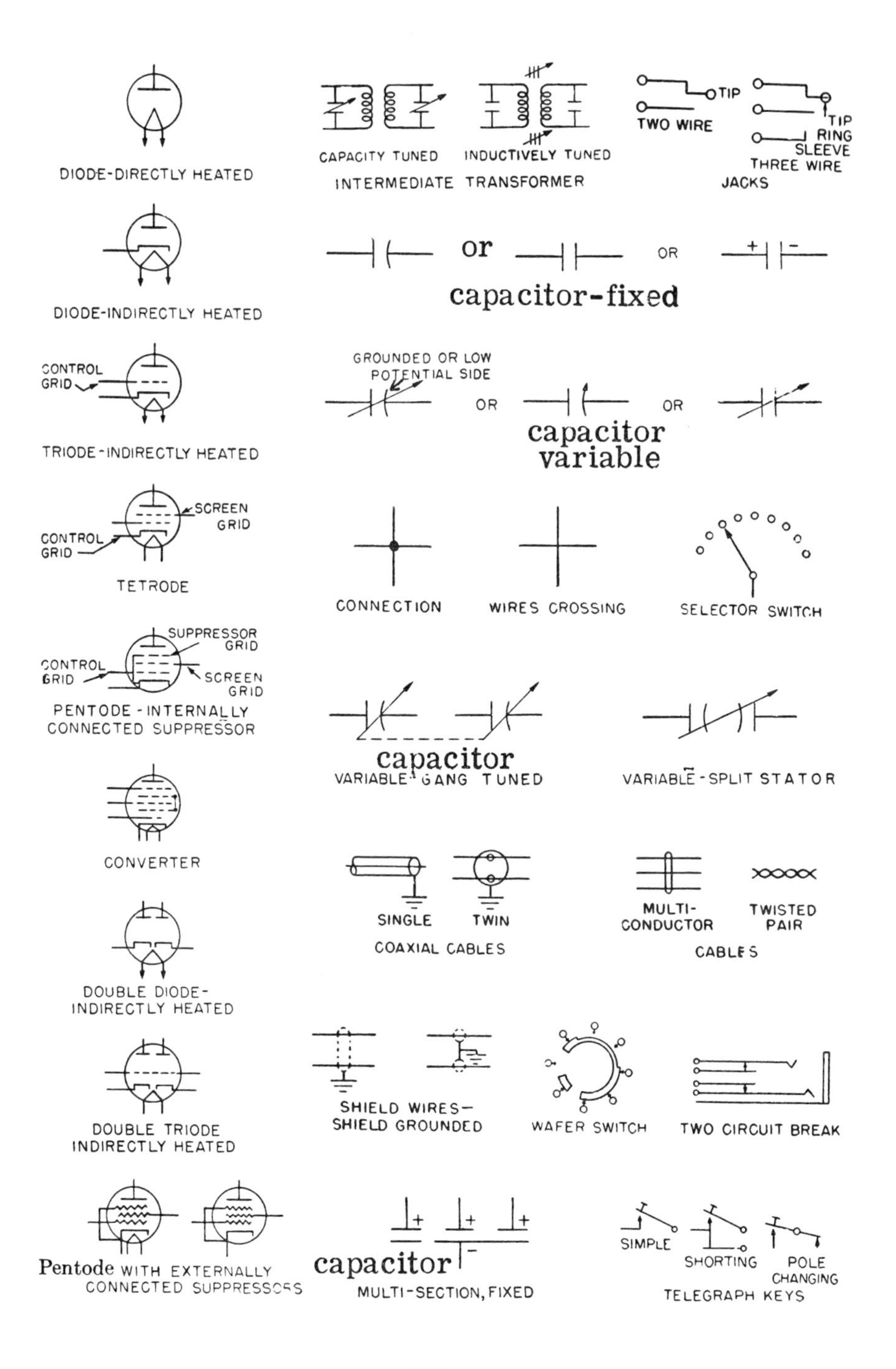

APPENDIX 3
RADIO ABBREVIATIONS

Group	Abbreviation	Meaning
Ampere	a, or amp.	ampere
	µa	microampere
	ma.	milliampere
Farad	fd or f	farad (rarely used alone)
	µf	microfarad
	µµf or pf	micromicrofarad or picofarad
Frequency	f	frequency
	c*(or).........	cycles
	cps	cycles per second
	kc.	kilocycles per second
	Mc.	Megacycles per second
Henry	h	henry
	mh	millihenry
	µh	microhenry
Impedance	X_L	inductive reactance (in ohms)
	X_C	capacitive reactance (in ohms)
Ohm	Ω (Omega)...	ohm resistance
	MΩ...........	megohm (one million ohms)
Volt	v	volt
Watt	w	watt
	p	power (in watts)
Current	AC	alternating current
	DC	direct current
Frequency	AF	audio frequency
	RF	radio frequency
	IF	intermediate frequency
	TRF	tuned radio frequency
Miscellaneous .	CW	continuous wave
	AM	amplitude modulation
	FM	frequency modulation
	EMF	electromotive force (in volts)
	MOPA	master oscillator power amplifier
	EST	Eastern Standard Time
	GMT	Greenwich Mean Time

*The term "Hertz" has been used in place of cycles in recent years. The abbreviation for Hertz is Hz. We can, therefore, also use the terms kiloHertz and MegaHertz.

The abbreviation for kiloHertz is kHz.
The abbreviation for MegaHertz is MHz.

APPENDIX 4
ANSWERS TO PRACTICE QUESTIONS

LESSON 1

1. c 2. b 3. b 4. c 5. b 6. c 7. a 8. b
9. 14W 10. 100V 11. 6.67Ω 12. b 13. c 14. b 15. a
16. c 17. b 18. c 19. a 20. d

LESSON 2

1. b 2. c 3. b 4. c 5. b 6. c 7. b

LESSON 3

1. b 2. c 3. d 4. c 5. d 6. b 7. c 8. a
9. c 10. c 11. a

LESSON 4

1. b 2. d 3. a 4. d 5. b 6. a 7. c 8. b
9. a 10. a

LESSON 5

1. b 2. b 3. a 4. b 5. b 6. b

LESSON 6

1. c 2. a 3. d 4. b 5. d 6. c 7. b 8. d
9. c 10. d

LESSON 7

1. c 2. b 3. a 4. c 5. c 6. a 7. a 8. d
9. b 10. d

LESSON 8

1. a 2. c 3. c 4. a 5. b 6. b 7. a 8. c

LESSON 9

1. c 2. d 3. c 4. c 5. a 6. d 7. b 8. a
9. c 10. d 11. b 12. b 13. d 14. a 15. a 16. a
17. d

LESSON 10

1. c 2. c 3. b 4. c 5. b 6. c

LESSON 11

1. a 2. a 3. a 4. c 5. b 6. b 7. c 8. b
9. a 10. b 11. c 12. c 13. a 14. c

LESSON 12

1. b 2. c 3. a 4. c 5. d 6. b 7. d

LESSON 13

1. d 2. c 3. b 4. b 5. b 6. c 7. d 8. c
9. b 10. c 11. d 12. d

LESSON 14

1. d 2. c 3. c 4. b 5. b 6. c 7. a 8. d
9. a 10. c 11. c 12. a 13. c 14. b

FINAL NOVICE CLASS FCC-TYPE EXAM

1. b 2. c 3. d 4. b 5. b 6. d 7. b 8. d
9. a 10. d 11. d 12. a 13. c 14. d 15. a 16. b
17. c 18. c 19. b 20. d 21. c 22. a 23. b 24. a

FREQUENCY ALLOCATIONS FOR POPULAR AMATEUR BANDS

All in MegaHertz. "X" indicates no privileges.

CLASSES	NOVICE		TECHNICIAN		GENERAL AND CONDITIONAL		ADVANCED		EXTRA	
BANDS	CW	PHONE	CW	PHONE	CW	PHONE	CW	PHONE	CW	PHONE
80 METERS	3.7 to 3.75	X	3.7 to 3.75	X	3.525 to 3.750 and 3.85 to 4.0	3.85 to 4.0	3.525 to 3.750 and 3.775 to 4.0	3.775 to 4.0	3.5 to 4.0	3.75 to 4.0
40 METERS	7.1 to 7.15	X	7.1 to 7.15	X	7.025 to 7.150 and 7.225 to 7.3	7.225 to 7.3	7.025 to 7.3	7.15 to 7.3	7.0 to 7.3	7.15 to 7.3
20 METERS	X	X	X	X	14.025 to 14.15 and 14.225 to 14.35	14.225 to 14.35	14.025 to 14.15 and 14.175 to 14.35	14.175 to 14.35	14.0 to 14.35	14.15 to 14.35
15 METERS	21.1 to 21.2	X	21.1 to 21.2	X	21.025 to 21.20 and 21.30 to 21.450	21.3 to 21.450	21.025 to 21.20 and 21.225 to 21.45	21.225 to 21.450	21.0 to 21.450	21.2 to 21.45
10 METERS	28.1 to 28.2	X	28.1 to 28.2	X	28.0 to 29.7	28.3 to 29.7	28.0 to 29.7	28.3 to 29.7	28.0 to 29.7	28.3 to 29.7
6 METERS	X	X	50.0 to 54.0	50.1 to 54.0	50.0 to 54.0	50.1 to 54.0	50.0 to 54.0	50.1 to 54.0	50.0 to 54.0	50.1 to 54.0
2 METERS	X	X	144.0 to 148.0	144.1 to 148.0	144.0 to 148.0	144.1 to 148.0	144.0 to 148.0	144.1 to 148.0	144.0 to 148.0	144.1 to 148.0

INDEX

N, O

P

Q, R

AMECO NOVICE CLASS ADDENDUM

This addendum is to be used with Ameco #7-01, 23-01 and 102-01 books. The following information, together with the material in the above listed books, will satisfy the FCC Novice examination requirements of PR Bulletin 1035A of August 31, 1983. See Page 8 for the availability of Ameco Novice Exams. The following facts should be thoroughly mastered.

RULES AND REGULATIONS

(1) A frequency band is a segment or group of adjacent frequencies in the radio frequency spectrum. For instance, the frequencies between 3,500 kHz and 4,000 kHz can be designated as a BAND of frequencies between 3,500 kHz and 4,000 kHz. We can also refer to these frequencies as the 80 meter BAND because 80 meters is the wavelength of the center frequency (3,750 kHz) of the band.
(2) Frequency privilege means the right to operate in a particular band of frequencies.
(3) The only code that a Novice class operator may use is the International Morse code.
(4) The term "emission" means the radio frequency energy coming from the antenna of a radio station.
(5) Unidentified radio communications or signals are emissions from a station without any radio identification from the operator sending them.
(6) To maliciously interfere means to willfully or deliberately interfere with unlawful intentions.
(7) The Amateur call sign consists of three parts: (1) the prefix, which can be one or two letters, (2) a single digit (Ø through 9), and (3) a suffix of one, two or three letters. There are four groups of call signs: Group A, Group B, Group C and Group D. Each group differs from the other in its structure. For instance, Group A has a one or two letter prefix, a digit and a one or two letter suffix. Group B has a two letter prefix, a single digit and a two letter suffix. The Novice class operators receive their calls from Group D. The single digit indicates the geographical location of the station.
(8) An Amateur may communicate with a station in a foreign country, unless the administration of that country or our country, has served notice that it objects to such communications.
(9) An Amateur station shall be identified by the transmission of its call sign at the end of each communication, and every ten minutes or less during a communication. The word "communication" is understood to mean a completed conversation. During a single communication, two amateurs can switch back and forth many times. Identification of a station should be given in plain English or International Morse code.
(10) Effective August 29, 1983, the following rules concerning maximum authorized transmitter power went into effect:

(a) "Notwithstanding other limitations, Amateur radio stations shall use the minimum transmitting power necessary to carry out the desired communications".

(b) Each Amateur radio station may be operated with a peak envelope output (transmitter power) not exceeding 1,500 watts, except as provided in paragraph (e) of this section". An operator is no longer required to provide a means for accurately measuring transmitter power. Operators are allowed to determine, individually, the best means for insuring compliance with the power limitations. The FCC has indicated a standard that it will recognize in measuring output. They suggest, to their own personnel and to amateurs, that the power output can be determined while the station is operating, as indicated by: (1) the reading of a thru-line peak reading radio frequency wattmeter, properly matched, or (2) calculation of the power using peak RF voltage, as indicated by an oscilloscope or other peak reading device.

(d) The peak envelope power output (transmitter power) of each amateur radio transmitter, shall not exceed 200 watts when transmitting in any of the following frequency bands: (1) 3,700-3,750 kHz. (2) 7,100-7,150 kHz (7,050-7,075 kHz when the terrestrial location of the station is not within Region 2, but is in Region 1 or 3); (3) 21,100-21,200 kHz; (4) 28,100-28,200 kHz.

(e) "An Amateur radio station may transmit A3 emissions on or before June 1, 1990, with a transmitter power exceeding that authorized by paragraph (b) of this section, provided that the power input (both radio frequency and direct current) to the final amplifying stage supplying radio frequency power to the antenna feedline, does not exceed 1,000 watts exclusive of power for heating the cathodes of vacuum tubes. The limitations of paragraphs (a) and (d) of this section will apply". This last rule is actually a "grandfather" period for those who use type A3 emissions. However, it has nothing to do with Novice class operators or others who are operating in the Novice sub-bands.

(11) If an Amateur receives an official notice of violation concerning the physical or electrical characteristics of his transmitting apparatus, he must answer this notice and state fully what steps were taken to prevent future violations, and if any new apparatus is to be installed, the date such apparatus was ordered, the name of the manufacturer, and promised date of delivery. If the notice relates to some lack of attention to, or improper operation of the transmitter, the name of the operator in charge shall be given.

OPERATING PROCEDURES

(1) An operator should send a CQ at the speed that he can receive CW. The other station will generally answer a CQ at the speed of the CQ. If an Amateur sends CQ at a speed higher than he can copy, he will have difficulty when the other station comes back to him at the speed of the CQ.

(2) When an operator "zero-beats" his signal with another station,

he knows that he is on the same frequency as the other station. This makes for less interference and allows more radio stations to use the band. When an operator hears CQ, he quickly zero-beats his transmitter with that of the CQ sender. He does this because he knows that the person sending the CQ will first listen on his own frequency. Therefore, he has a better chance of making the contact.
(3) A dummy antenna should be used in place of the actual antenna during testing and tuneup procedures. A dummy antenna is a resistance load that presents the same resistance and power dissipation to the final stage as the antenna does. The use of the dummy antenna reduces the amount of time that a transmitter is on the air during tuneup procedures.

RADIO WAVE PROPAGATION

(1) Ionospheric propagation is the propagation of a sky wave that leaves the antenna and travels skyward until it strikes the ionosphere. Here it is reflected (technically, the term should be refracted) back to earth some distance away. The terms, "sky wave propagation" and "skip propagation" mean the same thing as ionospheric propagation.
(2) Ground wave propagation refers to the radio wave that travels along the surface of the earth, gradually losing its strength until it is completely attenuated.

AMATEUR RADIO PRACTICES

(1) All antenna and rotor cables should be grounded when an Amateur radio station is not in use. This protects the radio station from damage due to lightning strikes.
(2) When the radio station is not in use, or during electrical storms, the AC power plugs of the equipment should be disconnected from their outlets. This will protect the station equipment from damage due to lightning striking the electrical wiring in the building.
(3) All metal cabinets housing the station equipment and accessories, should be connected directly to a common point with large diameter, short length conductors. Another heavy conductor should connect this common point to a good ground, such as the cold water pipe and/or a good external ground. The conductors that ground the equipment should be size #10 or larger.
(4) A horizontal wire antenna should be high enough so that a person on the ground cannot touch it.
(5) A person on the ground, assisting another person on an antenna tower, should wear a helmet (hard hat) to prevent head injuries in the event that any tools or parts are accidentally dropped.
(6) Receiver overload, or "front-end" overload, refers to a strong fundamental signal from a transmitter that gets into a receiver's front end. Because of its strength, it swamps the receiver's input circuits, which just don't have the selectivity to suppress the strong signal. RF interference to a receiver that is caused by front-end overload, generally interferes over a wide range of frequencies. In the case of a TV receiver, it causes interference to most or all channels. The

first step in curing this problem would be to install a high-pass filter at the TV receiver's input.
(7) A multiband antenna, connected to an improperly tuned transmitter, may radiate harmonic interference. This is because a multiband antenna is designed to operate at several frequencies that are harmonically related.
(8) Interference due to harmonic radiation from a transmitter, generally affects one or two channels of a TV receiver. This type of interference is eliminated by installing a low-pass filter at the transmitter.
(9) The impedance of the transmitter final amplifier circuit must match (be equal to) the impedance of the antenna or feedline, in order to transfer maximum RF power from the transmitter to the antenna or feedline. Matching impedances also cuts down on the radiation of spurious and harmonic signals. If a feedline (transmission line) is used, it is most important for the transmission line impedance to be equal to the impedance of the antenna at its feedpoint.
(10) An SWR bridge (called a reflectometer) reads forward and reflected voltages, and can show the SWR on the face of the meter.
(11) In order to determine the impedance match between a transmitter and an antenna, an SWR bridge should theoretically be connected between the transmission line and the antenna.
(12) Coaxial cable has an impedance of from 50 to 75 ohms, and matches the center impedance of a half-wave dipole. The outer conductor of the coaxial cable acts as a shield and prevents radiation from the cable. It also prevents interaction between the cable and nearby metal objects.
(13) A high SWR reading may indicate poor electrical contact between the parts of an antenna system.
(14) If the length of an antenna is wrong for the frequency of the signal being transmitted, it will cause a high SWR reading. This can be corrected by changing the length of the antenna.

ELECTRICAL PRINCIPLES

(1) The pressure in a water pipe can be compared to the electromotive force or voltage in an electrical circuit.
(2) A voltage has a positive (+) and a negative (-) polarity.
(3) The lowest voltage that will cause a current in an insulator is known as the Breakdown Voltage.
(4) A microampere is one millionth of an ampere.

CIRCUIT COMPONENTS

(1) The fundamental frequency of a quartz crystal depends primarily upon its thickness. The thinner the crystal, the higher is its fundamental frequency; the thicker the crystal, the lower is its fundamental frequency.
(2) The frequency stability of a crystal-controlled transmitter is generally better than that of a transmitter using a variable frequency oscillator.

(3) The figure below represents the schematic of a tetrode vacuum tube.

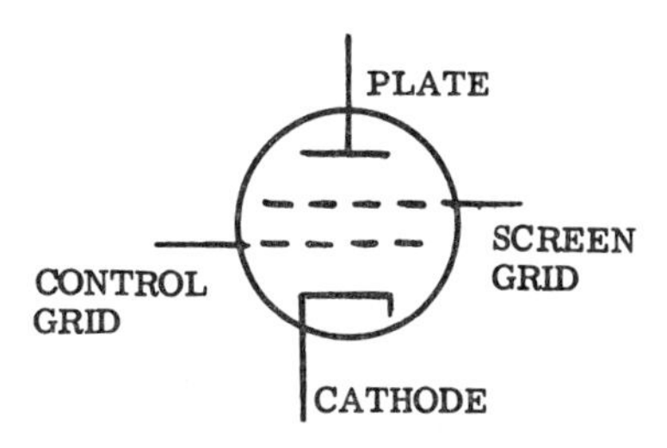

PRACTICAL CIRCUITS

(1) The figure below is a block diagram of a portion of an Amateur radio station including a transmitter, an antenna feedline, an antenna and an SWR bridge.

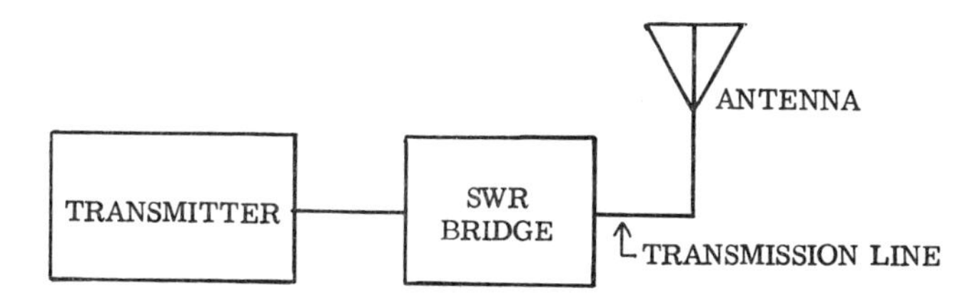

(2) The figure below represents a block diagram of a portion of an Amateur radio station including a transmitter, a receiver, a T-R switch, an antenna feedline and an antenna.

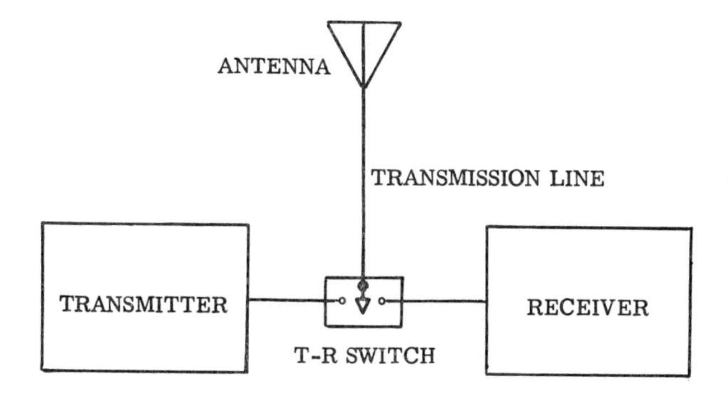

(3) The figure below is a block diagram of a portion of an Amateur radio station including a transmitter, an antenna tuner, an antenna feedline, and an antenna.

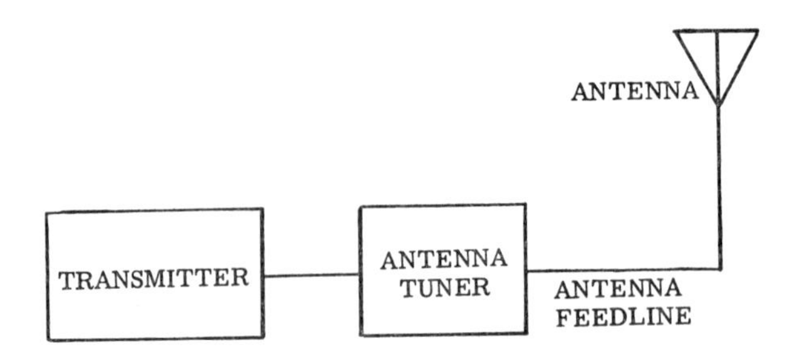

(4) The figure below is a block diagram of a portion of an Amateur radio station including a transmitter, a telegraph key, an antenna feedline and an antenna.

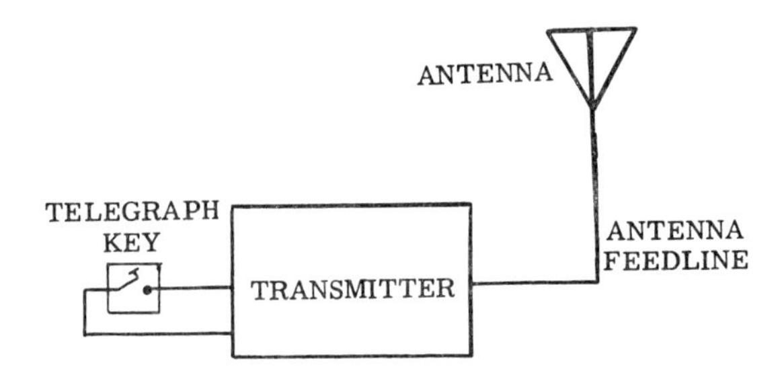

(5) The figure below is a block diagram of a portion of an Amateur radio station showing how two antennas and a dummy antenna can be switched to the same transmitter.

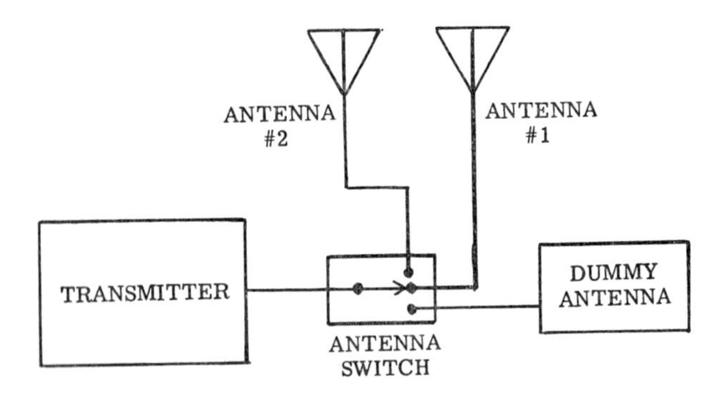

(6) The figure below is a block diagram of a portion of an Amateur radio station including a transmitter, an SWR meter, an antenna tuner, an antenna feedline and an antenna.

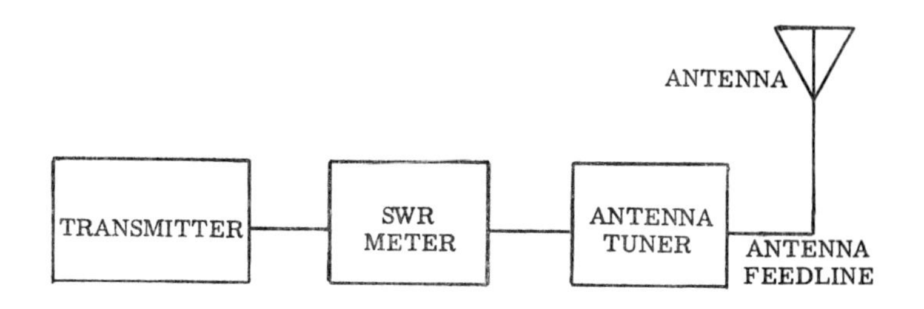

(7) The figure below is a block diagram of a typical Novice station including a transmitter, a receiver, an antenna feedline, an antenna, a T-R switch, grounding provisions, and a telegraph key.

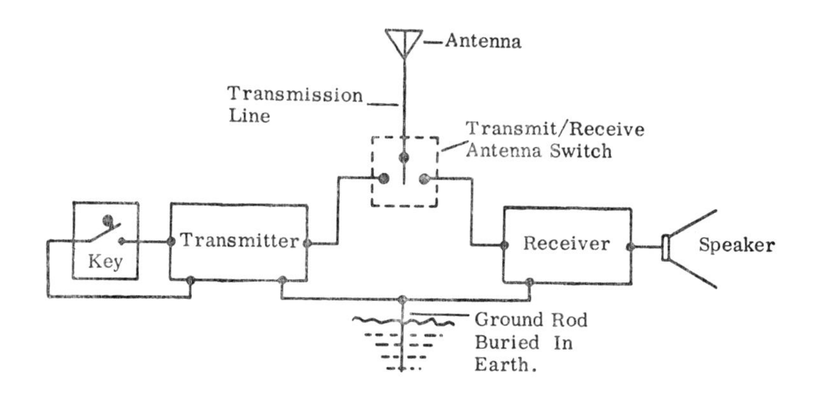

SIGNALS AND EMISSIONS

(1) Backwave may be caused by: incomplete neutralization, or keying of the final stage only, or inductive pickup between the antenna coupling coils and one of the lower power stages.
(2) Spurious emissions from a transmitter may be caused by improper neutralization of the power amplifier stage. They may also be caused by improper parts placement, improper lead dress, lack of proper shielding or lack of proper grounding.

ANTENNAS AND FEEDLINES

(1) The resonant frequency of an antenna, vertical or horizontal, varies inversely with its length. As the antenna length is increased, its resonant frequency is reduced, and vice versa.
(2) Some advantages of coaxial cable are: the outer wire braid acts as a shield, preventing spurious and harmonic radiation from the transmission line, it is simple to install, it has minimum noise pick-up, it is efficient. Coaxial cable can be buried directly in the ground for some distance without adverse effects because of its weatherproof vinyl covering. Coaxial cable can be located near grounded metal

objects because its outer conductor acts as a shield.
(3) Some advantages of a parallel conductor feedline are: (1) it is a balanced line and it is easier to match to a dipole antenna, which is also balanced; (2) the open wire type has a characteristic impedance of 200 to 600 ohms, and can be used to match high impedance antennas; (3) it has very low losses if it is properly constructed and properly used.

Some disadvantages of parallel conductor feedline are: (1) it is not shielded and can radiate or be affected by nearby metal objects; (2) it may be difficult to install because the wires can get tangled and touch each other; (3) the twin lead type exhibits very high losses in wet weather.